COHERENT COOPERATIVE PHENOMENA

KOGERENTNYE KOOPERATIVNYE YAVLENIYA

КОГЕРЕНТНЫЕ КООПЕРАТИВНЫЕ ЯВЛЕНИЯ

The Lebedev Physics Institute Series

Editors: Academicians D. V. Skobel'tsyn and N. G. Basov

P. N. Lebedev Physics Institute, Academy of Sciences of the USSR

Recent Volumes in this Series

Volume 35	Electronic and Vibrational Spectra of Molecules
Volume 36	Photodisintegration of Nuclei in the Giant Resonance Region
Volume 37	Electrical and Optical Properties of Semiconductors
Volume 38	Wideband Cruciform Radio Telescope Research
Volume 39	Optical Studies in Liquids and Solids
Volume 40	Experimental Physics: Methods and Apparatus
Volume 41	The Nucleon Compton Effect at Low and Medium Energies
Volume 42	Electronics in Experimental Physics
Volume 43	Nonlinear Optics
Volume 44	Nuclear Physics and Interaction of Particles with Matter
Volume 45	Programming and Computer Techniques in Experimental Physics
Volume 46	Cosmic Rays and Nuclear Interactions at High Energies
Volume 47	Radio Astronomy: Instruments and Observations
Volume 48	Surface Properties of Semiconductors and Dynamics of Ionic Crystals
Volume 49	Quantum Electronics and Paramagnetic Resonance
Volume 50	Electroluminescence
Volume 51	Physics of Atomic Collisions
Volume 52	Quantum Electronics in Lasers and Masers, Part 2
Volume 53	Studies in Nuclear Physics
Volume 54	Photomesic and Photonuclear Reactions and Investigation Method with Synchrotrons
Volume 55	Optical Properties of Metals and Intermolecular Interactions
Volume 56	Physical Processes in Lasers
Volume 57	Theory of Interaction of Elementary Particles at High Energies
Volume 58	Investigations in Nonlinear Optics and Hyperacoustics
Volume 59	Luminescence and Nonlinear Optics
Volume 60	Spectroscopy of Laser Crystals with Ionic Structure
Volume 61	Theory of Plasmas
Volume 62	Methods in Stellar Atmosphere and Interplanetary Plasma Research
Volume 63	Nuclear Reactions and Interaction of Neutrons and Matter
Volume 64	Primary Cosmic Radiation
Volume 65	Stellarators
Volume 66	Theory of Collective Particle Acceleration and Relativistic Electron Beam Emission
Volume 67	Physical Investigations in Strong Magnetic Fields
Volume 68	Radiative Recombination in Semiconducting Crystals
Volume 69	Nuclear Reactions and Charged-Particle Accelerators
Volume 70	Group-Theoretical Methods in Physics
Volume 71	Photonuclear and Photomesic Processes
Volume 72	Physical Acoustics and Optics: Molecular Scattering of Light; Propagation of Hypersound; Metal Optics
Volume 73	Microwave–Plasma Interactions
Volume 74	Neutral Current Sheets in Plasmas
Volume 75	Optical Properties of Semiconductors
Volume 76	Lasers and Their Applications
Volume 78	Research in Molecular Laser Plasmas
Volume 77	Radio, Submillimeter, and X-Ray Telescopes
Volume 79	Luminescence Centers in Crystals
Volume 80	Synchrotron Radiation
Volume 81	Pulse Gas-Discharge Atomic and Molecular Lasers
Volume 82	Electronic Characteristics and Electron–Phonon Interaction in Superconducting Metals and Alloys
Volume 83	Theoretical Problems in the Spectroscopy and Gas Dynamics of Lasers
Volume 84	Temporal Characteristics of Laser Pulses and Interaction of Laser Radiation with Matter
Volume 85	High-Power Lasers and Laser Plasmas
Volume 86	Superconductivity
Volume 87	Coherent Cooperative Phenomena
Volume 88	Cosmic Rays in the Stratosphere and in Near Space
Volume 89	Electrical and Optical Properties of III-IV Semiconductors

Proceedings (Trudy) of the P. N. Lebedev Physics Institute

Volume 87

Coherent Cooperative Phenomena

Edited by
N. G. Basov

P.N. Lebedev Physics Institute
Academy of Sciences of the USSR
Moscow, USSR

Translated from Russian by
Donald H. McNeill

CONSULTANTS BUREAU
NEW YORK AND LONDON

Library of Congress Cataloging in Publication Data

Main entry under title:

Coherent cooperative phenomena.

(Proceedings (Trudy) of the P. N. Lebedev Physics Institute; v. 87)
Translation of Kogerentnye kooperativnye i͡avlenii͡a.
Includes index.
1. Coherence (Optics)–Addresses, essays, lectures. 2. Quantum optics–Addresses, essays, lectures. I. Basov, Nikolaĭ Gennadievich, 1922- II. Series: Akademii͡a nauk SSSR. Fizicheskiĭ institut. Proceedings; v. 87.
QC1.A4114 vol. 87 [QC403] 530′.08s [535′.13]
ISBN 978-1-4757-0057-2 ISBN 978-1-4757-0055-8 (eBook)
DOI 10.1007/978-1-4757-0055-8 78-2008

The original Russian text was published by Nauka Press in Moscow in 1976 for the Academy of Sciences of the USSR as Volume 87 of the Proceedings of the P. N. Lebedev Physics Institute. This translation is published under an agreement with the Copyright Agency of the USSR (VAAP).

PREFACE

This anthology is devoted to the theoretical and experimental study of coherent phenomena in quantum optics. Considerable attention is devoted to the investigation of cooperative effects in multilevel molecular and atomic systems. The theoretical analysis is based on a single approach, using the theory of groups, which makes it possible to present the results in a compact and physically elegant form. The results of a detailed experimental and theoretical investigation of the photon echo, self-induced transparency, and nutation are presented. The theory of a coherent resonant mechanism for the interaction of laser radiation with multiatomic molecules which causes "prompt" dissociation is presented. A number of applied possibilities, in particular, the control of radiation by laser light and laser isotope separation, are analyzed.

CONTENTS

COHERENT PHYSICS AND ITS APPLICATIONS

L. A. Shelepin

A review is given of the current state of the theory of coherent cooperative phenomena. The basic methods of describing them are discussed, including the theory of the density matrix, the Dicke method and its generalizations, coherent states, the differential phenomenological approach, and statistical methods; their internal unity is demonstrated. The problems of kinetics are briefly discussed. A review of the basic applied approaches is given and their prospects are evaluated. The global character of coherent phenomena, in that they involve problems from widely varying areas of physics, is pointed out.

INTRODUCTION

This anthology is devoted to the physics of coherent cooperative phenomena. This new area of research has developed into an entire discipline in recent years. Two principal approaches are involved. The first deals with coherent electromagnetic radiation. At present high-power coherent radiation can be generated, its properties and interaction with material are being studied, and numerous practical applications are being devised. Laser technology is being developed rapidly. The second approach deals with the development of concepts about coherent cooperative states of matter. Experimental and theoretical studies are being made of such phenomena as superradiance (caused by phasing of normally independent molecular radiators), superconductivity, superfluidity, and superscattering (scattering of a coherent beam of material particles). Here the prefix "super" is typical and indicates that this group of phenomena is unusual and nonstandard relative to previously existing physical concepts.

All of these outwardly different phenomena from various branches of physics have a common feature which may be called coherence (cooperative behavior, phasing, correlation) of radiation or matter. Coherence is associated with an object made up of many particles in which the particles are not considered individually, but, for some reason, act in unison and display properties differing from those of isolated particles. At present the theory of coherent phenomena is just being formulated and has a fragmented character. In a number of cases the terminology has not even been established. In addition there has been a rapid expansion in theoretical and experimental research, and a number of unexpected new results have been obtained. Furthermore, as it goes far beyond the limits of one or several branches of physics the theory of coherent phenomena takes on a general physical significance. It should be emphasized that quantum mechanical features appear in the macroscopic phenomena of coherent physics and any fully adequate theory of these phenomena must be developed solely on the basis of quantum mechanical concepts. The technical applications are also being extended rapidly and there are still more important prospects ahead.

Since there are differences in the published literature in the treatment of a number of questions about coherent phenomena, knowledge of which is required to discuss the specific problems examined in this anthology, we shall preface this collection with a brief review of the main approaches to and methods and applications of coherent physics. This makes it possible to place each of the articles [1-6] in perspective. The discussion is purely qualitative and the references only illustrate the basic ideas and do not claim completeness. The basic methods used in the theory of coherent phenomena are discussed in Sections 1-4 and applied, and systematic questions are discussed in Sections 5-8.

1. The Density Matrix Formalism. Group Theoretical Aspects

The study of coherence effects in the theory of oscillations already has a long history (see the book by Gorelik [7]). Renewed development of this topic was made possible by the development (based on the phenomenon of induced emission discovered by Einstein [8]) of the first quantum generators of coherent electromagnetic radiation (i.e., masers) by Basov, Prokhorov, and Townes in 1954 [9, 10]. We note that the possibility of obtaining negative absorption was considered even earlier [11]. Over the next 20 years there was a rapid expansion of research on masers and lasers. It should be noted in addition that the coherent radiation obtained from lasers is only one of the coherent forms of motion of matter. The essence of these forms of motion is that under certain conditions a system of "noninteracting" particles displays new (nonstatistical) properties associated, roughly speaking, with the synchronization of the phases of the ψ-functions of individual particles, atoms, or molecules.

An important feature of the coherent properties of matter is their generality and fundamentally quantum nature. The technique for describing these phenomena is based on the density matrix formalism. The density matrix (or statistical operator) ρ gives the most general description of a quantum mechanical system. It describes both the statistics of purely quantum states and the statistical characteristics associated with our ignorance of certain properties of the object (as also occurs in classical statistics).

The diagonal elements of the density matrix give the populations, and the nondiagonal elements ρ_{nm} give the correlations between states. The quantity

$$g_{nm}=\frac{\rho_{nm}}{(\rho_{nn}\rho_{mm})^{1/2}} \tag{1}$$

may be defined as a measure of the degree of coherence. As the density matrix yields the most complete description of a physical system, it serves as a starting point for the analysis of coherent effects. This applies to the density matrix of both matter and fields.

The matrix ρ is Hermitian, i.e.,

$$\rho_{nm}=\rho^{*}_{mn}, \tag{2}$$

positive definite, i.e.,

$$\rho_{kk}\geqslant 0, \tag{3}$$

and normalized, i.e.,

$$\mathrm{Sp}\,\rho=1. \tag{4}$$

The Hermitian matrix ρ may be reduced to diagonal form by the unitary transformation T as follows:

$$\rho_{j}\delta_{jj'}=\sum_{nm}T_{jn}\rho_{nm}T^{-1}_{mj'}. \tag{5}$$

The elements of the matrix ρ are bounded by the condition

$$\mathrm{Sp}\,(\rho^2)=\sum_{nm}|\rho_{nm}|^2\leqslant 1. \tag{6}$$

The average value of the physical quantities corresponding to an operator Q are given by

$$\langle Q\rangle = \mathrm{Sp}\,(Q\rho). \tag{7}$$

The time evolution of the density matrix is described by the equations

$$\frac{\partial\rho}{\partial t} = -\frac{i}{\hbar}\,[H,\ \rho], \tag{8}$$

$$\rho(t) = \exp\left(-\frac{i}{\hbar}Ht\right)\rho(0)\exp\left(\frac{i}{\hbar}Ht\right) \tag{9}$$

and is determined by the Hamiltonian of the physical system. The problem of the evolution of the density matrix has been discussed in detail in [12-16].

Pure quantum mechanical states, described by a wave function ψ, are a special case of the density matrix. Then the density matrix is written in the form of the projection operator

$$\mathscr{P}(x,\ x') = \psi(x)\,\psi^*(x'). \tag{10}$$

A general density matrix may be expanded over pure states (projection operators) as

$$\rho(xx') = \sum_k W_k\psi_k(x)\,\psi_k^*(x'), \tag{11}$$

where $\sum_k W_k = 1$ and $W_k \geqslant 0$.

We note that the projection operators of Eq. (9) (cf. [17-19]) obey the relation

$$\mathscr{P}^2 = \mathscr{P}. \tag{12}$$

In general, as can be seen from Eq. (11), states described by a density matrix include all states of ψ with some weight for each. For real physical objects it is extremely difficult to solve the system of equations for the density matrix in order to analyze coherent effects. Thus, it has become necessary to develop general theoretical methods which do not reduce to a direct solution of these equations. As in the case of atomic and nuclear spectroscopy and elementary particle physics, group theoretical methods are very effective. We now discuss briefly the group theoretical aspects of the density matrix formalism.

In the two-level case the density matrix

$$\rho = \begin{Vmatrix} \rho_{11} & \rho_{12} \\ \rho_{21} & \rho_{22} \end{Vmatrix} \tag{13}$$

only has three independent variables. According to conditions (4) and (2), which imply $\rho_{11} + \rho_{22} = 1$ and $\rho^* = \rho_T$, the matrix ρ may be written in the form

$$\rho = \begin{Vmatrix} 1+p_z & p_x - ip_y \\ p_x + ip_y & 1 - p_z \end{Vmatrix} = E + \sigma_x p_x + \sigma_y p_y + \sigma_z p_z, \tag{14}$$

where σ_i are the Pauli matrices and E is the unit matrix. In the three-level case conditions (4) and (2) yield $\rho_{11} + \rho_{22} + \rho_{33} = 1$, $\rho^* = \rho_T$. The density matrix then has eight independent variables and may be written in the form

$$\rho = E + H_1p_1 + H_2p_2 + E_{\alpha_1}p_{\alpha_1} + E_{\alpha_2}p_{\alpha_2} + E_{\alpha_1+\alpha_2}p_{\alpha_1+\alpha_2} + E_{-\alpha_1}p_{-\alpha_1} + E_{-\alpha_2}p_{-\alpha_2} + E_{-(\alpha_1+\alpha_2)}p_{-(\alpha_1+\alpha_2)}, \tag{15}$$

where H_i and E_i are the canonical generators of the SU_3 group in a Cartan–Weil basis. Another form of Eq. (15) is also possible:

$$\rho = \sum_{i,\,k=1}^{3} A_{ik}p_{ik}. \tag{16}$$

The relations between the generators A_{ik} and the canonical generators are given by

$$\begin{aligned} &A_{11}=H_1\sqrt{3}+H_2, \quad A_{12}=E_{\alpha_1}\sqrt{3}, \quad A_{13}=E_{\alpha_1+\alpha_2}\sqrt{3},\\ &A_{21}=E_{-\alpha_1}\sqrt{3}, \quad A_{22}=-H_1\sqrt{3}+H_2, \quad A_{23}=E_{\alpha_2}\sqrt{3},\\ &A_{31}=E_{-\alpha_1-\alpha_2}\sqrt{3}, \quad A_{32}=E_{-\alpha_2}\sqrt{3}, \quad A_{33}=-2H_2. \end{aligned} \tag{17}$$

These equations are valid for an arbitrary number of levels; however, A_{ik} must be taken to mean the matrices

$$(A_{ik})_{\mu\nu}=\delta_{i\nu}\delta_{k\mu}-{}^1/_3\delta_{ik}\delta_{\mu\nu},$$

which, in the case of four levels, for example, are expressed in terms of 15 operators of a Cartan–Weil basis and a unit matrix.

The time-evolution equation for the density matrix may be written in an invariant form. For example, in the two-level case, we have

$$ih\frac{\partial\rho_i}{\partial t}=\varepsilon_{ikl}D_k\rho_l, \tag{18}$$

where $D_1 = H_{12} + H_{21}$, $D_2 = H_{12} - H_{21}$, and $D_3 = H_{11} - H_{22}$.

Analogous equations may be written for the three-level and arbitrary n-level cases. This is a clearly unitary-invariant form which reflects the SU_n-symmetry of the density matrix for finite systems.

Matrices of finite group transformations may be successfully employed to analyze the coherent properties of a system of weakly interacting atoms. In this case the density matrix is the product of the density matrices describing the behavior of the individual atoms,

$$\rho=\rho^{(1)}\rho^{(2)}\ldots\rho^{(N)}, \tag{19}$$

and transforms according to the representation $[D(P, Q, \ldots)]^N$.

For two independent subsystems $H = H_1 + H_2$, $\rho = \rho_1\rho_2$, the commutator $[H, \rho] = 0$ and ρ may be represented in the form

$$\rho=\alpha e^{-\beta H}, \tag{20}$$

where α and β are parameters.

This formula for the density matrix of the grand canonical ensemble may be rewritten in the form

$$\rho=\exp\left(-q+\beta\sum_i N_i\mu_i-\beta H\right), \tag{21}$$

where μ_i is the chemical potential.

For a two-level system, $\rho_{11}=\alpha e^{\beta H_{11}}$ and $\rho_{22}=\alpha e^{-\beta H_{22}}$.

For $p_z = \frac{1}{2}(\rho_{11} - \rho_{22})$ and using $\rho_{11} + \rho_{22} = 1$, we have

$$p_z=i\tan\left[\frac{1}{2}\beta'(H_{22}-H_{11})\right]=i\tan\left(\frac{i\Delta E}{kT}\right), \tag{22}$$

where $H_{11} - H_{22} = \Delta E$ and $\beta' = -i/kT$.

The quantity $i\Delta E/kT$ may be related to the vector parameter for a finite rotation matrix [20].

In the general case of a mutilevel (multiparticle) system the use of finite transformations of the SU_n group also seems promising.

Noncompact groups will play an important role in the analysis of a relativistic density matrix as in the analysis of systems with an infinite number of states. Thus, a harmonic oscillator is described by representations of the SU(1:1) group [21]. Noncompact groups and infinite-dimensional representations (including an infinite number of states all at once) also appear in the study of such systems as the rotator or the hydrogen atom.

2. Dicke's Method and Its Generalizations

Dicke's work, published in 1954, has had a significant effect on research about coherent cooperative states of matter [22]. This theory deals with a complex system consisting of a large number of emitters and was used to study collective states of radiating molecules. An idealized model of a molecule with two energy levels was used. The states of such a molecule are described by representations of the SU_2 group; i.e., spin operators are used. The intensity I of coherent spontaneous emission, as opposed to the usual spontaneous emission intensity I_0, is determined by the half-difference of the populations m and by a certain cooperative number r which specifies the character of the correlation of the molecules; that is,

$$I = I_0 (r + m)(r - m + 1). \tag{23}$$

The quantum number r does not change during any radiative transitions of the system. Special bound states of the system in which the intensity is proportional to the square of the number of molecules, N^2 (r = 1/2, m = 0), the so-called superradiant states, as well as states in which the system cannot radiate at all (m = r + 1) are possible. We note that there is not an inverted population in the superradiant case, as opposed to lasers. The coherent radiation is spontaneous, rather than induced, and is produced by correlation of the emitters. Later Dicke's problem was generalized to multilevel systems [23-25]. This was done because of the need to consider the interaction of matter with intense radiation fluxes and because of the analytical situation which results when several resonant pulses interact.

The wave functions of an n-level system in energy space transform according to a representation of the SU_n group. The numbers which specify the signature of the representation are the cooperative characteristics and are conserved during all radiative transitions of a multilevel system. We emphasize that classification of the wave functions of an n-level system according to the SU_n group basically signifies an extension of Eqs. (14) and (15) and the unitary symmetry of the density matrix to pure states. A real system is described by the density matrix. Thus, any comparison of the density matrix method and Dicke's method is inadmissible. From this standpoint the introduction of group characteristics for the wave function (the "dichotomous" variable of [12]) is a convenient procedure resulting from the symmetry of the density matrix. On the whole, Dicke's method may be defined as a group theoretical method for the density matrix of finite systems.

We note that the generators of the SU_n group are expressed in terms of bilinear combinations of boson operators (cf. [3]). In turn, any physical quantities of a system with a fixed number of particles may be found with the aid of polynomial expansions in the generators and, therefore, of the second quantization operators.

3. The Method of Coherent States

A third important method of describing collective phenomena is the theory of coherent states. For many years the theory of coherence was developed only to study optical phenomena. Its task was to reveal the features of interference and diffraction in the field of randomly modulated electromagnetic waves generated by thermal light sources. The approach was purely classical and used correlation functions. With the development of quantum optics a number of unusual effects were observed, for example, high orders of coherence which did not reduce to an interference effect alone and even appeared when interference was absent. The basis of modern quantum optics is the coherent states of oscillators (discussed in detail by Glauber [26]). As opposed to the customary problems of quantum mechanics in which the states ψ corresponding to a definite energy are examined, coherent states correspond to a particle which is simultaneously in all levels of an oscillator. The development of the mathematical apparatus for and research on coherent states have led to the widespread introduction of the idea of coherent states in various branches of physics.

Coherent states are a definite superposition of an infinite number of wave functions corresponding to the states of a system with a certain number of particles $|n\rangle$. The coherent oscillator states which have been studied in most detail are those which are vectors of the form

$$|z\rangle = \exp\left(-\frac{1}{2}|z|^2\right)\sum_{n=0}^{\infty}\frac{z^n}{(n!)^{1/2}}|n\rangle, \tag{24}$$

valid for arbitrary complex numbers z. In terms of second quantization operators the notation for $|z\rangle$ has the form

$$|z\rangle = \exp\left(-\frac{1}{2}|z|^2\right)\sum_{n=0}^{\infty}\frac{(za^+)^n}{n!}|0\rangle. \tag{25}$$

Coherent states have a number of remarkable properties. Thus, the vector $|z\rangle$ is the eigenvector of the annihilation operator; i.e.,

$$a|z\rangle = z|z\rangle. \tag{26}$$

A coherent state is the ground state of the displacement operator,

$$D^{-1}(\beta)\,aD(\beta) = a + \beta, \qquad D^{-1}(\beta)\,a^+D(\beta) = a^+ + \beta^* \tag{27}$$

and of the translation operator

$$|z\rangle = D(z)|0\rangle. \tag{28}$$

While not orthogonal, they do have the completeness property. In addition the coherent states of an oscillator minimize the uncertainty relations and are the states closest to classical. Since they include the wave functions of all the levels, they make it possible to introduce the phase. These properties of the coherent oscillator states are the cause of their wide range of applications. The techniques of the theory of coherent states, originally developed for the requirements of quantum optics [26-28], have been successfully carried over to other branches of physics and, primarily, used to analyze certain boson systems (cf. [29]). The Glauber approach was used in [29-30] for the theory of superfluidity. Here the representation of the coherent states was associated with the function which serves as an ordering parameter in phenomenological descriptions of superfluidity. Coherent states have been introduced for a system of excitons in the Bose–Einstein condensation problem for excitons. The method of coherent states has also been used to describe the motion of a particle in a magnetic field and to compute the transition probabilities between the Landau levels [31], as well as in the Heisenberg model of ferromagnetism [32], for spin waves, in quantum electrodynamics to remove the infrared catastrophe, in elementary particle physics in connection with dual models [33], and in the analysis of superconductivity.

If the coherent states of oscillators have been discussed extensively and found a wide range of applications, then the attempt to generalize them to nonequidistant systems has encountered a number of difficulties. Coherent angular momentum states of the form

$$|z_1z_2\rangle = \exp\left[-\frac{1}{2}(|z_1|^2+|z_2|^2)\right]\sum_{R=0}^{\infty}\sum_{M=-R}^{R}\frac{z_1^{R+M}z_2^{R-M}}{[(R+M)!\,(R-M)!]^{1/2}}|RM\rangle \tag{29}$$

were constructed in [34] and coherent states of a rotator and a spherical top, in [35, 36]. The construction is based on a representation of the Hamiltonian in terms of boson operators, that is, on the oscillator variables. It is necessary to note that the concept of coherent states for nonequidistant systems undergoes a transformation. The correspondence with the classical case is no longer complete. Not all properties of the classical motion are preserved. Thus, the coherent states of a rotator ($H = J^2/2I$),

$$|z_1z_2t\rangle = [\cosh|z|^2]^{-1/2}\sum_{j=0}^{\infty}\sum_{M=-j}^{j}\frac{z_1^{j+M}z_2^{j-M}}{[(j+M)!\,(j-M)!]^{1/2}}|jM\rangle\exp\left[-\frac{i\hbar}{2I}j(j+1)t\right], \quad |z|^2=|z_1|^2+|z_2|^2, \tag{30}$$

become smeared out. During the time evolution of the system the coordinates of the rotator axis become indeterminate. However, for these states it is possible to introduce a conserved momentum vector which at any moment of time is limited to narrow regions in all directions.

Along with the derivation of specific generalizations of coherent states there have been attempts to construct a state of the Glauber type for arbitrary quantum systems. The method proposed by Malkin and Man'ko [37] for introducing these states is based on the construction of creation and annihilation operators from the initial coordinates of the system and their conjugate momenta. The coherent states are defined as the eigenstates of the annihilation operator which is an integral of the motion. Perelomov [38] has proposed a method of constructing coherent states for an arbitrary Lie group. The states constructed by him differ in structure from Eq. (29) for the SU_2 group and from the states given in [39] for noncompact groups. Hence, the problem of constructing coherent states is not solved uniquely in general, and this ambiguity is fundamental. All generalized coherent states may be represented in the form of infinite series in powers of the creation and annihilation operators similar to Eq. (25). Physically speaking, one such series is indistinguishable from the others. In principle, it is not impossible that various generalized coherent states may be realized for a single object. Thus, any comparison of them with one another is inadmissible.

All series of this type are directly related to the general theory of the density matrix for an infinite number of states. In addition, it would apparently be correct to take the behavior closest to classical as the basis of a generalization of the coherent states. In general we might expect the following properties to hold for such states [36]: their evolution is described by the classical equations of motion; the physical quantities in the equations of motion have small dispersion (are "almost exactly determined") although they may correspond to noncommutating operators; and the uncertainty relations are minimized.

Coherent states are the quantum states closest to classical and therefore determine the transition between classical and quantum mechanics. From this standpoint a classical state is closest to an infinitely weighted sum of quantum states, rather than to a single fixed state.

4. Phenomenological Approaches

In the analysis of coherent phenomena, quantum mechanics presently impinges on the traditional field of classical physics. Widely different branches of optics have been regarded as classical until recently, or classical models have been used successfully to describe the phenomena involved. Thus, for example, although quantum levels are used in the analysis of scattering processes, the dynamics of the processes itself is well described classically. The concept of phase arises in coherence techniques, and coherent states are closest to classical; thus, we have a sort of return to classical representations but with a quantum mechanical basis.

Thus, in general the classical and semiclassical theories, while using the language of phase relationships, have a phenomenological character. To a certain extent the work of Ter-Mikaelyan and Feinberg (cf. [40]) on cooperative phenomena occurring during the scattering of light in crystals touches on this.

Of special interest is the phenomenological approach associated with nonlinear differential equations. Nonlinear equations were first used by Ginzburg and Landau [41] to describe coherent processes in the theory of superconductivity. They introduced an ordering parameter ψ with the normalization condition $|\psi|^2 = n_s$, where n_s is the density of superconducting electrons. This parameter plays the role of an effective wave function for the superconducting electrons. The Ginzburg–Landau equations for ψ, the vector potential **A**, and the field **H** can be written in the form

$$\begin{gathered}\frac{1}{2m}\left(-i\hbar\nabla-\frac{2e\mathbf{A}}{c}\right)^2\psi+\alpha\psi+\beta|\psi|^2\psi=0,\\ \frac{\operatorname{curl}\mathbf{H}}{2\pi}=\frac{\mathbf{j}}{c}=\frac{e\hbar}{imc}(\psi^*\nabla\psi-\psi\nabla\psi^*)-\frac{4e^2}{mc^2}|\psi|^2\mathbf{A}.\end{gathered} \tag{31}$$

The coefficients α and β are taken from the microscopic theory.

These nonlinear equations (31) describe well a wide circle of phenomena associated with superconductivity. In recent years a number of general phenomena in the evolution of nonlinear waves in dispersive media described by the Korteweg–de Vries equation have been studied [42]. It has been shown that in general for sufficiently long times the initial perturbation decays into a set of stationary solitary waves (solitons) and into a nonsoliton part having the form of a wave packet which diffuses with time. The problems associated with the formation of solitons and their properties determined by the nonlinearity [43] are now of great interest in plasma physics, hydrodynamics, elementary particle physics, astrophysics, and biophysics. In addition, in various optical effects (for example, in self-induced transparency which occurs during the propagation of ultrashort pulses through a resonantly absorbing medium, [44, 45]) solitons are produced as a result of coherent processes. Thus, there is a definite relationship between coherent and nonlinear effects. There is great interest in the study of this relationship and in the attempt to understand the mechanism involved.

It is possible that in describing coherent phenomena with nonlinear differential equations we are dealing with a differential representation (or realization) of the group theoretical approach. In their article, Karasev and Shelepin [3] analyze two aspects (differential and difference) of the second quantization operators used to describe coherent phenomena and demonstrate the feasibility of using the finite-difference approach along with the differential approach.

Statistical methods may also play an important role in the analysis of coherent phenomena. Here the starting point is the fact that, in view of their closeness to the classical description, the coherent states must have a relationship to the Feynman path integral method [46]. Malkin and Man'ko [47] have studied this relationship and have shown that the construction of the coherent states of an arbitrary quantum system is equivalent to evaluating a path integral.

We note that path integrals were first introduced by Einstein and Smoluchowski in their work on the Brownian motion.

Subsequently, the Feynman formulation of quantum mechanics was introduced. In it the evolution operator U or Green function for the Schrödinger equation is described by an integral over a trajectory in coordinate or phase space.

In the analysis of coherent effects with a finite number of states, for example, in Dicke's method, the interrelation with certain sums over paths becomes important. In Dicke's method

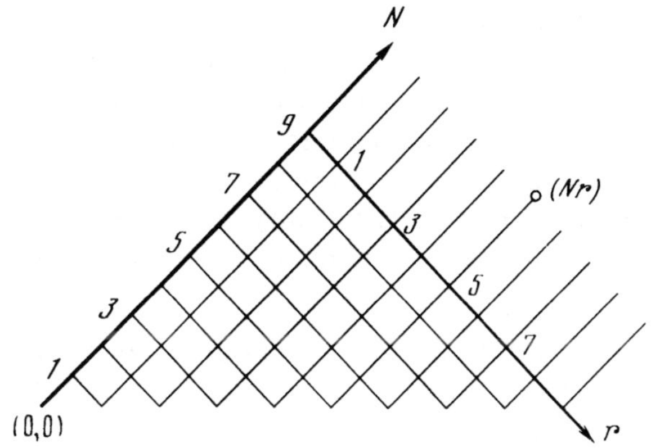

Fig. 1. A diagram of the paths from point (00) to point (Nr).

the coefficient A(N, r) in the expansion

$$[D(1/2)]^N = \sum_{r=0}^{N} A(N, r)\, D(r), \tag{32}$$

which denotes the multiplicity of the representation D(r), determines the probability of a given cooperative state in the presence of N two-level molecules. It may be shown that A(N, r) is specified by the number of trajectories from point (00) to point (Nr) of Fig. 1. Motion is only to the right (upward and downward) and along continuous lines. This number is equal to the difference of two binomial coefficients,

$$A(N, r) = C^r_{N+r} - C^{r-1}_{N+r}. \tag{33}$$

The problem of resonance excitation of multiatomic molecules (cf. the article by Akulin et al. [6]) also reduces largely to the analysis of certain sums of trajectories.

On the whole, the use of sums of trajectories (of integrals along trajectories) to analyze coherent phenomena is very promising both for finite and infinite numbers of states. In turn, there is a characteristic interrelation between summation of trajectories and the group theoretical approach, on the one hand, and the statistical methods (random walk problems, Monte Carlo methods), on the other.

From the standpoint of a general theory of coherent phenomena there is thus an internal unity among such approaches as the density matrix formalism, Dicke's method, the method of coherent states, the description by means of nonlinear differential equations, and the group theoretical and statistical analyses.

5. The Kinetics of Coherent Phenomena

In general, to study some process while taking coherent effects (coherent kinetics) into account it is necessary to solve complicated systems of kinetic equations which describe the real physical object. Along with the equations for the diagonal elements of the density matrix, which correspond to the populations, they include an equation for the nondiagonal elements, which correspond to the correlations among states. The problems associated with a general analysis of such systems have hardly been studied.

At the same time, population kinetics, as a special case of coherent kinetics, has been developed in some detail [48, 49]. Its development has been stimulated by the requirements of laser physics.

Laser action has been obtained on an immense number of specific transitions involving practically all types of atom (ion, molecule) in all states of aggregation (solid, liquid, gas, plasma) with very different forms of pumping (chemical, electrical, thermal, optical, nuclear).

A real substance, made up of atoms and molecules, has a complicated system of levels and terms. The population distribution in such an externally pumped system is determined by internal relaxation processes. As a result, a population inversion cannot possibly occur over those levels on which the pumping source acts. Thus, there was a demand for detailed development of new means for describing the nonequilibrium states of matter. A number of qualitative conclusions from population kinetics which retain their validity in the general case can point out ways of constructing a kinetic theory of coherent phenomena. We shall mention the most important aspects.

First, in a majority of cases it is impossible to limit oneself to two- or three-level schemes since real physical objects and multilevel systems must be considered. Thus, the two-level idealization of the Dicke method is applicable to a very limited number of systems in which there is a resonance interaction with the field and becomes inadequate for describing the interaction with intense radiant fluxes. Here there is a direct analogy with population kinetics. The initial approach to the problem of creating an inverted population was also to analyze two- or three-level schemes in which pumping to the upper level and emptying of the lower level were specified. These schemes were very limited and have been replaced by a multilevel approach which consists of constructing a model for the kinetics of a nonequilibrium substance, isolating the controlling processes, and solving the equations for the populations of many levels on a computer.

Second, it is necessary to note the presence of a hierarchy of times for various relaxation processes in a complex atomic or molecular system. This is characteristic of population kinetics [49]. In coherent kinetics there is a set of relaxation times for the nondiagonal elements T_2 (the transverse relaxation times) in addition to the set of relaxation times for the populations T_1 (the longitudinal relaxation times). Usually T_2 is much less than T_1; that is, the coherence of the substance is first destroyed, then the populations relax to their equilibrium states. The differences in these characteristic times makes it possible to study relaxation in a real system taking cooperative effects into account and to separate the dominant processes from those which play a second-order role in a given effect.

Finally, the possibility of extending the method of quasiequilibrium distribution functions developed recently for population kinetics [48] to coherent kinetics must be pointed out. Quasiequilibrium distributions usually depend on a small number of parameters which are functions of time. As opposed to the equilibrium distribution (Boltzmann, Bose, Fermi) they are valid only for certain degrees of freedom (groups of levels) and exist for a definite limited time during a relaxation process and are thus a characteristic of the process. Quasistationary distributions have been obtained for atomic and molecular levels for multiplicities of ions, for induced Compton scattering components, and for spin systems. The characteristic features of such distributions are clearly visible in the example of the Treanor distribution [50] for the the level populations x_n of an anharmonic oscillator,

$$x_n = x_0 \exp(\gamma n) \exp(-E_n/T). \tag{34}$$

It is distinguished from a Boltzmann distribution by the factor $\exp(\gamma n)$ and depends on two parameters, the gas temperature T and the parameter γ. The parameter γ can be related to the total energy stored in an anharmonic oscillator. If this energy exceeds the equilibrium amount then the upper levels are overpopulated relative to a Boltzmann distribution. Under nonequilibrium conditions without additional pumping of energy it is possible to achieve a redistribution of the vibrational energy by changing only the gas temperature. For mixtures of molecules under certain conditions the distribution functions also depend strongly on the concentrations of the molecular components. Thus, by changing the vibrational energy distribution with the aid of such "rudders" as the gas temperature and component concentrations it is possible to control the radiation within certain limits. The general technique for finding quasiequilibrium distributions to describe population relaxation reduces to making a series expansion in the time derivatives of the population of a given level [48]. In a sense this is equivalent to expanding in powers of the displacement of the system from equilibrium. The equilibrium distribution functions are the zeroth approximation. The first approximation is a quasiequilibrium distributions of the sort in Eq. (34). In principle this procedure may be carried over to coherent kinetics. This would make it possible to reduce the complicated system of equations for the general case to some equations for a small number of parameters which determine the quasiequilibrium distribution. On the basis of a study of the prospects for influencing these parameters it might be possible to control various coherent processes. A number of problems in multilevel kinetics will be discussed in this book [1, 2, 6]. Among earlier work on this topic we may note that of Askaryan [51].

6. Applied Problems. Photon Technology

In any analysis of applied problems in coherent physics a number of fundamental aspects will arise. In this section we shall mainly limit ourselves to a discussion of the applications of coherent radiation. Many types of laser have been developed. Lasing has been obtained over a wide range of wavelengths from the millimeter and submillimeter bands to the vacuum ultraviolet, and the possibility of lasing in the x-ray region is being discussed. Peak pulsed powers of 10^{12} W and cw powers of 10^5 W have been obtained. The practical applications of lasers are based on the coherent properties of laser radiation: high signal purity with frequency stability of order 10^{-14}, sharp directionality of the beam with negligible divergence so the radiation can be focused to a region of size of the order of the wavelength, immense electric fields (up to 10^7 V/cm) produced when the beam is focused, ease of controlling the beam, and high brightness temperature of the source. Numerous technological applications have been developed on the basis of these properties.

In industry lasers are used for metal cutting, cutting patterns in leather, sheet metal welding, microscopic welding and welding inside sealed vessels, punching holes, and surface machining. In medicine lasers find application in reattaching detached retinas, puncturing the eyeball to reduce the inner eye pressure in glaucoma, removing skin tumors, and in sealing blood vessels. Military uses include laser range finders and laser gyroscopes, guidance systems, and laser weapons (in 1971 a report appeared about an aircraft's having been shot down by a laser due to destruction of its aerodynamic quality [52]). In atmospheric and space research the applications are associated with laser ranging, measuring the distances to astronomical objects, tracking systems, and studies of the level of air pollution. The remarkable properties of coherent radiation make lasers an irreplaceable instrument for scientific research. Research is under way to initiate thermonuclear reactions with lasers. These are only some of the areas in which lasers are now used. It is to be expected that the applications of lasers will become far more numerous.

The operational question is, however, much broader. The problem is the fundamental level at which the energy of the coherent radiation is being used. The principal motive force first used by mankind was muscle power of people and domestic animals, then steam machinery became dominant, and, since the beginning of the twentieth century, electricity. Power plants and the production of electrical energy now substantially characterize the potential of a country. In the near future it is expected that an immense amount of energy will be obtained in the form of coherent radiation. The use of a costly (from the standpoint of negative entropy) form of energy for conversion to heat, as is done at present, is irrational. What is needed is both the effective conversion of "coherent" energy to other forms of energy and its direct utilization to do work.

The statement of the problem is now related to the cost of energy. Thus, the quantitative difference in the efficiency of steam and electrical machinery and in the economy of energy transmission over a distance leads to a sharp qualitative difference in the level of technology. From this standpoint the difference between the use of coherent radiative energy for heating (and the subsequent use of the thermal energy) and its direct use is a fundamental one.

The question is thus: Is it possible in principle to displace electrical technology from its dominant position by photon technology. While an analog of an electrical generator, the laser, already exists and is being developed, the analog of the electric motor, a photon machine, does not yet exist. It must be built. Christiansen and Hertzberg [53] have analyzed the properties and design of a photon machine using the concrete example of a closed cycle gas-dynamic laser consisting of a reservoir of hot compressed gas, a nozzle, a laser cavity, a diffuser, and a compressor. As the heated gas expands through the nozzle its temperature falls, the various vibrational modes have different relaxation rates, a divergence develops between the vibrational temperatures, and an inversion occurs – lasing takes place in the laser cavity. The vibrational energy, however, is not completely used up as radiation. As the gas is compressed in the diffuser the vibrational energy is converted to translational, the energy of directed motion of the particles goes into heat, and the mixture is heated and compressed. Little work is then required for the compressor to pump the gas back to the original cavity.

As in the case of a generator and motor, the laser and photon machine may switch roles. Then an absorbing section replaces the laser section and the lower level is more populated than the upper. On absorbing the coherent radiant energy the gas gains additional vibrational energy, partly due to absorption and partly due to cooling of the gas during absorption (cf. [54]). The gas pressure at the diffuser outlet exceeds the initial pressure and the compressor becomes a turbine. Thus, in the case of an ideal photon machine the coherent energy is absorbed and work is done. This system is one of the possible photon analogs of the electrical generator–motor system. In the design of such systems the problems of obtaining coherent light from a nonequilibrium state of matter and of the interaction of light with matter are two aspects of one and the same problem. It probably also will be important to utilize the coherent properties of matter in building photon machines. Along with the key problem of constructing photon "motors," some other aspects should also be noted.

An important advantage of photon technology is the possibility of wireless energy transfer. Because of the smallness of the diffractive divergence which causes expansion of the laser beam, coherent optical energy may be transported along a straight line for many thousands of kilometers. This is important under earthbound conditions and vital in outer space. A laser beam aimed from the earth illuminates only a small area, of the order of a hundred meters, on the moon. By delivering energy in such a manner with photon machines it might be possible to adjust the energy supply to the moon and other planets, to supply rockets, etc.

Engines using the energy of coherent radiation are of considerable interest. Thus, in principle, a laser may be used as an engine for a photon rocket with a maximal material outflow velocity [55]. Already the possibility of using lasers for maneuvering spaceships in orbit is being discussed from a practical point of view. This may be achieved both by direct ejection of photons and by ejection of material particles which flow into space due to interaction with the laser. By using a laser to vaporize matter from objects in space without engines it is possible to move them in the required direction (cf. [56]).

Coherent optics is also of fundamental import in the area of information recording. In holographic recording all the information about the object contained in the light is used (i.e., the amplitude and phase) as opposed, for example, to photography which requires only the amplitude. A holographic image creates a complete illusion of the presence of the object, and the object may be "looked at" from different sides. The image is practically indistinguishable from the real object. The capacity of a hologram is immense – an amount of information equal to 10^7 book pages may be recorded on a 100-cm^2 plate.

The use of the energy of coherent radiation in specific technologies is governed by its transformation into other forms of energy (thermal, mechanical, chemical, electrical, triggering thermonuclear reactions, etc.). In the near term the most realistic possibility is the construction of efficient electrical converters of light energy and the stimulation of chemical reactions with lasers. A number of new prospects have been opened up for the stimulation of chemical reactions in recent years. It has been shown that under certain nonequilibrium conditions quasiequilibrium population distributions develop, leading to a large overpopulation of the upper vibrational levels, and the dissociation rate may be greatly increased compared to the thermal rate. This makes it possible to obtain large amounts of free atoms (or radicals) and to produce chain reactions [49]. The reactions can be controlled by acting on the parameters of these distributions. For diatomic molecules energy pumped into a resonance transition does not cause efficient breaking of the bond (in this case the bonds in other molecules which are in the gas may be broken). Such diatomic molecules may, however, serve as an efficient quasistatic "reservoir." By pumping such reservoirs and choosing an appropriate mix of molecules it is possible to stimulate chemical reactions even without using the associated resonance radiation every time and (thus) with available fixed-frequency lasers. New possibilities for stimulating chemical reactions with collisionless processes during the interaction of intense laser radiation with matter have now been discovered (see [6]).

There is great interest in the problem of stimulating a thermonuclear reaction with lasers. On the other hand, there is the possibility of a laser reactor [57] in which nuclear energy is directly converted into coherent radiant energy, and in principle the efficiency may be greater than when heat is obtained directly.

7. Some Possible Uses of Coherent Phenomena

In the previous section we have examined applications involving the use of coherent radiation produced by lasers. Evidently, equally promising ideas should be discovered when coherent states of matter are used.

Already coherence, the quantum-mechanical theory of particle correlations, and the cooperative properties of multiparticle systems are important in optics, solid state physics, superconductivity and superfluidity, and nuclear physics. Meanwhile, questions about the practical use of the coherent states of matter are still mainly fragmentary. Thus, besides such developed areas of application as superconductivity, a number of applied possibilities are only beginning to be discussed or are in a formative stage. We now briefly consider the most distinct trends.

During the 1960's the experimental study of collective effects in the interaction of radiation with matter began. Here we may list induced self-transparency, photon echo, and nutation effects caused by the "coherentization" of radiators over times shorter than the relaxation time. Self-transmission of a pulse occurs because the leading edge of the pulse perturbs the medium and a correlation develops among the radiators (which is described by the nondiagonal elements of the density matrix). An ultrashort pulse takes a form such that its passage does not change the populations in the material. If we neglect relaxation processes, the pulse itself does not change either – that portion which is absorbed in the leading edge of the pulse is coherently reemitted in the trailing edge. To produce a photon echo two laser sources are needed. The second pulse, falling on a medium which has been made coherent by the first, causes a third pulse, the echo. To observe the nutation effect, which is an optical analog of striking a rotating gyroscope, a laser pulse is made to act on a sample together with a background from a steady-state laser: Then nutational vibrations are recorded in the emission from the sample. A number of molecular constants which describe the relaxation processes have been determined on the basis of these experiments. There is also the possibility of using the self-transmission effect for communication under upper atmospheric conditions. These effects and their applications are discussed in detail in this anthology [4, 5].

Considering multiple levels in discussions of the interaction of radiation with molecules opens a series of new possibilities beyond those for two-level systems. Superradiant states may occur simultaneously on several transitions. Two pulses at different frequencies corresponding to transitions between the levels of a molecule may interact coherently. This kind of indirect interaction between two light beams through matter takes place even with three levels. The radiation can be controlled using the third level in two ways: A pulse propagating in the medium may be modulated with another pulse, perhaps of much lower power, or the operating regime of the laser may be changed by a supplementary light pulse without changing the laser parameters.

We note that the role of the third level may be compared to that of the grid in a triode; that is, here are the elements of a unique kind of "supermicroelectronics" involving the interaction of radiation and elementary oscillators (atoms and molecules). This important new property, radiation control, is considered in more detail in the article [2] by Kudrya et al. in this anthology. The problem of ultrashort pulses borders on this group of problems. Generally in nonstationary fast processes the particles can no longer be regarded as completely independent. Due to perturbation of the medium a correlation develops among the particles and is preserved over times shorter than the relaxation times; that is, during a short interaction the system of particles responds as a unit. Ultrashort pulses are themselves formed as a result of coherent processes. When a bleachable (under the action of light) filter is present in a laser an effect known as self-synchronization of modes (self-mode locking) takes place [58]. This effect amounts to phasing of standing light waves (modes) belonging to different frequencies and lying within the spectral width of the line. Ultrashort pulses of duration 10^{-11}–10^{-12} sec may then be formed. Phasing of the waves occurs due to coherent process in the filter. The correlation among the modes means that they cannot be regarded as independent. Thus, we have a special sort of coherence for waves with different frequencies. The practical importance of ultrashort pulses is obvious. They may be used, for example, to measure short relaxation times (10^{-11}–10^{-12} sec), to initiate thermonuclear reactions, and to obtain very strong electric fields.

Coherent effects may be important in scattering as well as in absorption and emission. These effects are described by higher-order perturbation theory and may appear in the phenomenon of superscattering, where when the emitters are phased the scattering may be proportional to N^2. These questions are discussed in detail by Kudrya et al. [2]. Scattering on an initially excited medium is known as active spectroscopy and has become one of the most important research topics in nonlinear optics. Active spectroscopy makes it possible to determine radiative transition probabilities and transverse and longitudinal relaxation times, and to obtain induced combination scattering on a forbidden transition. Two pulses are used in active spectroscopy: the exciting pulse which transforms the medium to a state described by a nonstationary density matrix with nondiagonal elements in the energy representation and the probe pulse which scatters coherently on this medium.

In addition, a cooperative effect may be produced under real combination (or Rayleigh) scattering conditions due to perturbation of the medium when the leading front of the light pulse passes through it [59]. The existence of coherent cooperative effects in induced combination scattering was predicted by Sushchinskii [60] on the basis of experimental studies of specific angular and energy distributions which did not fit the usual theory.

Besides the use of short pulses to produce coherent states in matter, there is great interest in the production of coherent beams of particles which interact anomalously with matter (in this case the cross section is not proportional to the number of particles in the beam, N, as usual, but to N^2). The coherence of the particles in such a beam may be caused by nonstationary processes over times less than the relaxation times.

Up to now we have examined coherent states associated with a nonequilibrium state of matter. However, superradiant states may be realized under equilibrium conditions as well. In a molecule–radiation system there is a phase transition of the second type as the temperature is reduced and the state of the radiators is collectivized [61]. This is a direct analogy with the coherent state of superconductors and with superfluidity which also occur as a result of a phase transition. In the case of equilibrium superradiant states it is already impossible to speak of particles and radiation separately, and an overall coherent state of radiation and matter is formed which is energetically more favorable at low temperatures.

Clearly, these states may be used to produce short pulses.

The practical significance of superconductivity and superfluidity is generally known. There has been special interest in the possibility of high-temperature superconductivity (presently the maximum critical temperature for a superconducting transition is 23.2°K [62]).

It should be noted that fundamentally new uses for coherent mechanisms may arise in addition to those already available. In their article Akulin et al. [6] analyze a coherent resonant mechanism for the interaction of laser radiation with matter involving transitions among a series of almost equidistant levels. Such a mechanism is possible due to nonresonant absorption in the laser field. For a sufficient density of levels a multilevel equidistant system (or even a quasicontinuous background) will be formed. Prompt dissociation (and ionization) may take place by this mechanism. It might be very efficient in controlling chemical reactions and separating isotopes, and also opens the possibility of studying biological objects.

From this discussion it is clear that the essence of applied coherent physics is not only in the separate, individual applications, although they are important, but in the general fundamental transformations as well.

8. Technical Problems Associated with Coherence

In addition to applied problems in coherent cooperative phenomena, there is great interest in a number of questions of general physical methodology. A detailed analysis of them will evidently be associated to some extent with the further development of the theory. We now list the most important (in our opinion) problems:

1. The general universal nature of coherent phenomena. They enter modern physics at all three levels of the structure of matter (atomic, nuclear, elementary particles). We note, for example, the use of coherent techniques in the theory of weak and electromagnetic interactions and in the theory of the spontaneous breakdown of vacuum symmetry. The observation of common fundamental features in seemingly widely different, diverse fields of physics emphasizes the inner unity of matter and points the way to the interrelation between coherent phenomena and the structure of matter (and with the hierarchy of the structure of matter). The question also arises of creating a coherent gravitational field and controlling gravity.

2. The special properties of coherent interactions. These problems include ordering and negative entropy, correlation and force-free interaction problems, and interactions with coherent particles. When a coherent state is present it is already impossible to speak of individual particles (or quasiparticles) and the system may be considered only as a whole. In cases of simultaneous coherent states of radiation and matter it is impossible to speak of the existence of particles and radiation as such.

3. The relation of coherent states to symmetry. Examples of this relation are the descriptions of the cooperative states of two-level systems with the SU_2 group, of multilevel nonequidistant systems with the SU_n group, and of oscillator systems with the noncompact SU(1, 1) group. A systematic analysis is needed of the common symmetry of coherent phenomena. Associated with this is the problem of examining a coherent assembly of particles

statistically. The occurrence of cooperative properties and correlations among particles may substantially change the statistical features of an assembly of particles. An attempt [63] to introduce parastatistics in the description of coherent effects is indicative of this change.

4. The nature of coherent effects. This, in particular, includes studies of the foundations of quantum mechanics and of the relation between the wave-particle duality and cooperative phenomena, and analyses of the relation between classical and quantum theory on the basis of coherent states.

5. The unity of widely differing properties of physical, mathematical, and statistical descriptions of coherent effects. This is an indication of the prospects for substantial generalization of existing concepts.

Therefore, coherent physics now combines various physical approaches, requires complicated solutions, a certain coordination of widely different areas of physics (optics, solid state physics, nuclear physics, elementary particle physics, etc.), and the concentration of effort on promising problems. Research on coherent radiation is now of great practical significance. New technological applications must also be realized. The analysis of these questions about the coherent state of matter requires further scientific preparation, the solution of a number of fundamental theoretical problems, and the carrying out of a broad program of experimental research.

LITERATURE CITED

1. I. M. Lizin, T. M. Makhviladze, and L. A. Shelepin, Trudy FIAN, 87:21 (present issue).
2. V. P. Kudrya, T. M. Makhviladze, I. G. Sinitsyn, and L. A. Shelepin, Trudy FIAN, 87:38 (present issue).
3. V. P. Karasev and L. A. Shelepin, Trudy FIAN, 87:55 (present issue).
4. S. S. Alimpiev, Trudy FIAN, 87:92 (present issue).
5. S. S. Alimpiev, N. V. Karlov, and B. G. Sartakov, Trudy FIAN, 87:134 (present issue).
6. V. M. Akulin, S. S. Alimpiev, N. V. Karlov, and L. A. Shelepin, Trudy FIAN, 87:141 (present issue).
7. G. S. Gorelik, Oscillations and Waves [in Russian], GIFML, Moscow (1959).
8. A. Einstein, Phys. Z., 18:121 (1917).
9. N. G. Basov and A. M. Prokhorov, Zh. Éksp. Teor. Fiz., 27:431 (1954).
10. J. P. Gordon, H. J. Zeiger, and C. H. Townes, Phys. Rev., 95:282 (1954).
11. V. A. Fabrikant, Trudy Vys. Énerg. Inst., No. 41, p. 273 (1940).
12. V. M. Fain, Photons and Nonlinear Media [in Russian], Sov. Radio, Moscow (1972).
13. F. Berten, Foundations of Quantum Electronics [Russian translation], Mir, Moscow (1971).
14. F. Arecchi, M. Scully, H. Haken, and W. Waidlich, Quantum Fluctuations in Laser Radiation [Russian translation], Mir, Moscow (1974).
15. M. Lax, Fluctuations and Coherent Phenomena [Russian translation], Mir, Moscow (1974).
16. D. N. Zubarev, Nonequilibrium Statistical Thermodynamics [in Russian], Nauka, Moscow (1971).
17. R. M. Asherova and Yu. F. Smirnov, Nucl. Phys., B4:399 (1968); A114:116 (1970).
18. F. I. Fedorov, Zh. Éksp. Teor. Fiz., 35:493 (1958).
19. A. A. Bogush and F. I. Fedorov, Izv. Akad. Nauk BSSR, Ser. Fiz.-Tekh., 3:26 (1962).
20. E. E. Tkharev and F. I. Fedorov, Yad. Fiz., 5:1112 (1967).
21. I. M. Lizin, T. M. Makhviladze, and L. A. Shelepin, Izv. Akad. Nauk SSSR, Ser. Fiz., 37:2224 (1973).
22. R. H. Dicke, Phys. Rev., 93:99 (1954).
23. L. A. Shelepin, Zh. Éksp. Teor. Fiz., 54:1463 (1968).
24. E. Sh. Teplitskii, Teor. Mat. Fiz., 2:399 (1970).

25. T. M. Makhviladze and L. A. Shelepin, Zh. Éksp. Teor. Fiz., 62:2066 (1972).
26. R. Glauber, in: Quantum Optics and Quantum Electronics [Russian translation], Mir, Moscow (1966).
27. E. Wolf and L. Mandel, Rev. Mod. Phys., 37:231 (1965).
28. D. Klauder and E. Sudarshan, Fundamentals of Quantum Optics, Benjamin (1968).
29. A. Casher and M. Revzen, Ann. Phys., 48:441 (1968).
30. I. S. Langer, Phys. Rev., 167:183 (1968).
31. I. A. Malkin and V. I. Man'ko, Zh. Éksp. Teor. Fiz., 55:1014 (1968); Teor. Mat. Fiz., 6:71 (1971).
32. K. H. Donylass, Ann. Phys., 62:383 (1971).
33. A. N. Kvinikidze, S. D. Popov, D. T. Stoyanov, and A. N. Tavkhelidze, Teor. Mat. Fiz., 6:166 (1971).
34. R. Bonifacio, D. M. Kim, and M. O. Scully, Phys. Rev., 187:441 (1969).
35. R. W. Atkins and I. C. Dobson, Proc. Roy. Soc. London, A321:371 (1971).
36. T. M. Makhviladze and L. A. Shelepin, Yad. Fiz., 15:1081 (1972).
37. I. A. Malkin and V. I. Man'ko, FIAN Preprint No. 15 (1971).
38. A. M. Perelomov, Commun. Math. Phys., 26:222 (1972).
39. A. O. Barut and L. Girardello, Commun. Math. Phys., 21:41 (1971).
40. M. L. Ter-Mikaelyan, The Effect of Media on Electromagnetic Processes at High Energies [in Russian], Izd. AN ArmSSR (1969).
41. V. L. Ginzburg and L. D. Landau, Zh. Éksp. Teor. Fiz., 20:1064 (1950).
42. N. J. Zabusky and M. D. Kruskal, Phys. Rev. Lett., 15:240 (1965); V. I. Karpman and V. P. Sokolov, Zh. Éksp. Teor. Fiz., 54:1568 (1968).
43. B. B. Kadomtsev and V. I. Karpman, Usp. Fiz. Nauk, 103:193 (1971).
44. S. L. McCall and E. L. Hahn, Phys. Rev., 183:457 (1969).
45. I. A. Poluektov, Yu. M. Popov, and V. S. Roitberg, Usp. Fiz. Nauk, 114:97 (1974).
46. R. Feynman and A. Hibbs, Quantum Mechanics and Path Integrals, McGraw-Hill (1965).
47. I. A. Malkin and V. I. Man'ko, Zh. Éksp. Teor. Fiz., 59:1746 (1970); V. I. Man'ko, in: Coherent States in Quantum Theory [Russian translation], Mir, Moscow (1972), p. 5.
48. Theoretical Problems in the Spectroscopy and Gas Dynamics of Lasers, Trudy FIAN, Vol. 83, Nauka, Moscow (1975).
49. B. F. Gordiets, A. I. Osipov, E. V. Stupochenko, and L. A. Shelepin, Usp. Fiz. Nauk, 108:655 (1972).
50. C. E. Treanor, J. W. Rich, and R. G. Rehm, J. Chem. Phys., 48:1798 (1968).
51. G. A. Askar'yan, Zh. Éksp. Teor. Fiz., 46:403 (1964); 48:666 (1965).
52. Aviation Week and Space Technology, 12:6 (1970).
53. W. H. Christiansen and A. Hertzberg, Proc. IEEE, 61:1050 (1973).
54. B. F. Gordiets, A. I. Osipov, and R. V. Khokhlov, Zh. Éksp. Teor. Fiz., 44:1063 (1974).
55. L. I. Gudzenko and L. A. Shelepin, Kosmich. Issled., 3:167 (1962).
56. G. A. Askar'yan and E. M. Moroz, Zh. Éksp. Teor. Fiz., 43:2319 (1962).
57. L. I. Gudzenko and S. I. Yakovlenko, Krat. Soobshch. Fiz., No. 2, p. 14 (1974).
58. A. J. de Maria, D. A. Stetser, and H. Haynan, Appl. Phys. Lett., B, p. 174 (1966).
59. T. M. Makhviladze and L. A. Shelepin, Phys. Rev., A9:538 (1974).
60. M. M. Sushchinskii, Krat. Soobshch. Fiz., No. 2, p. 3 (1972).
61. K. Hepp and E. H. Lieb, Ann. Phys., 76:360 (1973).
62. J. R. Gavaler, Appl. Phys. Lett., 23:480 (1973); L. R. Testardi, J. H. Wernick, and W. A. Royer, Sol. State Commun., 15:1 (1974).
63. G. S. Kruglik, Preprint Phys. Inst. Byelorussian Acad. Sci., Minsk (1970).

SUPERRADIANCE EFFECTS IN MOLECULAR SYSTEMS

I. M. Lizin, T. M. Makhviladze, and L. A. Shelepin

A systematic generalization of Dicke's theory of coherent spontaneous emission to multilevel radiators and second-order perturbation theory is presented. Superradiant effects and the structure of the coherent responses of an equilibrium system to pulses (for equidistant and nonequidistant spectra) are examined. Coherent spontaneous scattering effect are discussed.

INTRODUCTION

Intense light pulses of approximate duration 10^{-12}-10^{-13} sec are now obtained with lasers. This makes it possible to observe effects caused by the coherent interaction of light with matter in practically any medium. The condition for an interaction to be coherent is that the pulse length be short compared to the characteristic damping time T_2 of the polarization induced in the medium by the field (the transverse irreversible relaxation time). Some possible effects which permit study of important spectral and relaxation characteristics of resonant media are the photon echo [1, 2], self-induced transparency [3], and so on.

In a coherent interaction individual atoms or molecules no longer radiate independently. For example, in the effects mentioned above an indirect interaction among the particles through the overall radiation field leads to so-called superradiant states which radiate at an intensity greatly exceeding the level from independent molecules. The phenomenon of superradiance was first discussed by Dicke [4] who showed that because of an interaction through a field two-level molecules can behave as a single quantum mechanical system whose radiation is determined by its coherent characteristics.

In recent years various aspects of Dicke's problem have been studied by many authors; however, these discussions have seldom gone beyond the two-level idealization and first-order perturbation theory. Meanwhile, the two-level idealization is a satisfactory approximation of a real system only when the interaction with the field is resonant in nature. It is important to include the multilevel character of the system in a number of practically important situations such as the following: when several resonant pulses at different frequencies act on a medium in the case of molecules with an equidistant spectrum; when describing an interaction with intense radiant fluxes, where it is necessary to use higher-order perturbation theory to evaluate combination scattering or harmonic generation effects; etc. Hence it is necessary to eliminate the above limitations in theoretical studies of superradiance effects. It should be noted that the approach based on a generalized Dicke theory is not the only possible one for this group of problems. Nevertheless, it is of undoubted pedagogical value.

Up to now the generalized Dicke problem has been discussed in a small number of papers. In [5] it was interpreted from the standpoint of permutation groups. In [6, 7] the possibility of a direct generalization of the energy spin method to a system of multilevel radiations was considered. Methods of calculating the coherent responses of matter to the action of several pulses using a matrix generalization of the Dicke formula (Eq. (72) in [4]) were developed in [8]. Consideration of the cooperative behavior of a system of radiators in the framework of Dicke's theory has been limited up to now to spontaneous emission alone, that is, to first-order processes, except in [9, 10]. Equations for the scattered intensity as a beam of photons passes through a medium were obtained in [9]. In these equations cooperative effects were taken into account by terms quadratic in the density of scattering particles. Walls [10] has obtained the scattering probabilities for a system of molecules in a Dicke state with the maximum cooperation number.

This article is devoted to a discussion of a systematic generalization of the Dicke theory and is based on our previous work [11-14]. A multilevel description is analyzed. Here the mathematical apparatus of the theory of group representations is used to permit examination of all the problems mentioned in [15]. Cooperative effects in the second-order perturbation theory are discussed in Sections 2 and 3. These effects are important in studies of the interaction of matter with powerful short light pulses when relaxation processes are unable to return the medium to its initial equilibrium state. In a perturbed medium, scattering on individual molecules may be correlated; that is, the molecules may behave as a single system. The appearance of nondiagonal elements in the density matrix of the medium (energy representation) leads to a coherence effect caused by a number of peculiarities of scattering processes under these conditions. In Section 2 the probabilities of spontaneous Stokes and antistokes scattering on pure cooperative states are derived. In Section 3 the probabilities of spontaneous excitation pulse of duration $\ll T_2$ are calculated. Superradiance effects in equidistant systems are discussed in Section 4.

1. Superradiance Effects in a System of Multilevel Molecules (Nonequidistant Spectrum)

We shall examine the characteristic features of the emission from nonequidistant molecules using the example of a system of radiators with three nondegenerate energy levels. We shall assume that the linear dimensions of the volume occupied by the particle ensemble are small compared to the wavelengths emitted by the system and neglect inelastic collisions (a low density gas). Then the internal energy variables and the coordinates of the centers of mass are separable. The Hamiltonian of the system can be written using the generators of the SU_3 group; see, for example, [15, 16]. In the following we shall use the notation introduced there. The internal energy operator for the j-th molecule has eigenvalues $\varepsilon_1 = -\varepsilon/2$, $\varepsilon_2 = \varepsilon/2$, and ε_3 and we shall write it in the form

$$H_1^{(j)} = \varepsilon I_3^{(j)} + \varepsilon_3 \left(\frac{1}{3} - Y\right)^{(j)}. \tag{1}$$

The eigenfunctions of this operator are $\chi_i^{(j)}$, i = 1, 2, 3, and transform as the D(1, 0) representation of the SU_3 group. The unperturbed Hamiltonian of a system of N molecules has the form

Here U_g ($\mathbf{r}_1, \mathbf{r}_2, \ldots, \mathbf{r}_N$) is the position-dependent part of the wave function ($H_0 U_g = \varepsilon_g U_g$); the functions $\chi^{(j)}$ show at which level each of the molecules is to be found; and N_1, N_2, N_3 are the level populations. The eigenvalues of I_3 and $1/3 - Y$ are equal to $1/2(N_2 - N_1)$ and $1/3\,(N_1 + N_2 - 2N_3)$, respectively.

The interaction of molecules with a classical radiation field is described by the operator

$$H_{\text{int}} = -\mathbf{A}\,[(\mathbf{e}_1 - i\mathbf{e}_2)\,I_+ + (\mathbf{e}_3 + i\mathbf{e}_4)\,K_- + (\mathbf{e}_5 + i\mathbf{e}_6)\,L_- + \text{H. conj.}] \tag{4}$$

in the dipole approximation, where $\mathbf{A}$ is the vector potential of the field at the system location and the vectors $\mathbf{e}_i$ are related to the matrix elements of the dipole moment by ($\hbar = c = 1$):

$$\mathbf{e}_1 - i\mathbf{e}_2 = -i\varepsilon\mathbf{d}_{12}; \qquad \mathbf{e}_3 - i\mathbf{e}_4 = i\,(\varepsilon_3 - \varepsilon/2)\,\mathbf{d}_{32}; \qquad \mathbf{e}_5 - i\mathbf{e}_6 = i\,(\varepsilon_3 + \varepsilon/2)\,\mathbf{d}_{31}.$$

Thus, the transition probabilities are determined by the matrix elements of the nondiagonal elements of the generators of the SU_3 group.

By analogy to the two-level case [4] it is appropriate to use linear combination of the states of Eq. (3) rather than these states alone so that H_{int} relates a single initial state to the least possible number of other states. Such a set of states is the basis of the irreducible representations $|P, Q, I, I_3\, Y\rangle$ of the SU_3 group.* Electromagnetic transitions conserve the quantum numbers P and Q. (The complete Hamiltonian is constructed linearly from generators of the group which commute with the Casimir operators.) It can be shown that for given N_1, N_2, N_3, and I the numbers $s \equiv 1/3(N - P + Q)$ and $a \equiv 1/3(N - P - 2Q)$ have values satisfying the inequalities

$$I_M - I \leqslant s \leqslant I_M + I, \qquad a \leqslant I_M - I, \qquad s + a \leqslant I_M - I + N_3, \tag{5}$$

where $2I_M = N_1 + N_2$ and $|N_2 - N_1| \leq 2I \leq N_1 + N_2$. In principle, these considerations make it possible to compute the transition probabilities for processes of arbitrary order in perturbation theory.

To first order in the perturbation theory the intensities of spontaneous emission from the state $|s, a, I, I_3, Y\rangle$ are

$$J_{i\to k} = J^{(0)}_{i\to k} f_{ik}\,(s,\ a,\ I,\ I_3,\ Y), \tag{6}$$

where $J^{(0)}_{ik}$ are the intensities of spontaneous emission of an isolated molecule in the $i \to k$ transition. The functions f_{ik} are determined by the matrix elements of the nondiagonal generators of the SU_3 group [12]:

$$\begin{aligned} f_{21} &= (I + I_3)\,(I - I_3 + 1), \\ f_{31} &= \gamma_1^2\,(I - I_3 + 1) + \gamma_2^2\,(I + I_3), \\ f_{32} &= \gamma_1^2\,(I + I_3 + 1) + \gamma_2^2\,(I - I_3), \end{aligned} \tag{7}$$

where

$$\begin{aligned} \gamma_1^2 &= (I_M + I - s + 1)\,(I_M + I - a + 2)\,(I_M + I - s - a + N_3)\,[(2I + 1)\,(2I + 2)]^{-1}, \\ \gamma_2^2 &= (I - I_M + s)\,(I_M - I - a + 1)\,(I_M + I - s - a + N_3 + 1)\,[2I\,(2I + 1)]^{-1}. \end{aligned} \tag{8}$$

Thus, the intensities depend on three quantum numbers, s, a, and I, one of which (I) changes in the $3 \to 1$ and $3 \to 2$ transitions. These numbers, which are the cooperation charac-

* Here and in the following the coordinate part of the wave function is omitted for brevity.

teristics of the molecular system, determine the rate of radiative decay of the system. In special cases when the number of particles in one of the levels is zero, the correlation between the other levels is described, as in the case of two-level molecules, by the appropriate isospin (I, K, or L). The emission in the other channels is proportional to the number of particles in the level from which the transition under consideration takes place. For example, if $N_2 = 0$, Eq. (5) implies $I = -I_3 = N_1/2$, $a = 0$, and $0 \leq s \leq \min(N_1, N_3)$. With Eq. (6) we obtain

$$J_{31} = J_{31}^{(0)}(L + L_3)(L - L_3 + 1), \quad J_{32} = J_{32}^{(0)} N_3, \quad J_{21} = 0, \tag{9}$$

where

$$L = \frac{1}{2}(N_1 + N_3) - s, \quad |L_3| \leqslant L \leqslant \frac{1}{2}(N_1 + N_3), \quad L_3 = \frac{1}{2}(N_3 - N_1).$$

For $s = 0$ [$L = (N_1 + N_3)/2$], $J_{31} = J_{31}^{(0)} N^2$ (the superradiant Dicke state). We made a similar analysis of Eq. (6) in [12, 15]. We now discuss the most interesting case of superradiance over all frequencies. For $I = I_M$ it follows from Eq. (6) that $a = 0$, and $0 \leq s \leq \min(N_1 + N_2, N_3)$. If $s = 0$, then

$$J_{i \to k} = J_{i \to k}^{(0)} N_i (N_k + 1), \quad \text{where} \quad i > k. \tag{10}$$

In particular, for equal populations the radiant intensity over all channels is proportional to the square of the total number of molecules. This case corresponds to complete symmetry of the wave function which transforms according to the D(N, 0) representation. Such states may be excited if we begin with the ground state, for which $I = -I_3 = N/2$ and $s = a = 0$.

After two pulses with frequencies ω_{12} and ω_{13} act on it the system goes into a superradiant state since during absorption of the first pulse the number I does not change and during absorption of the second pulse this number retains a maximum value. In this case the radiant intensity is calculated according to Eq. (10). After excitation of a system initially in thermodynamic equilibrium by two pulses, superradiance will also be observed [see Eqs. (14) with $\Gamma(\mathbf{k}_{12}, \mathbf{k}') = \Gamma(\mathbf{k}_{13}, \mathbf{k}''') = \Gamma(\mathbf{k}_{13} - \mathbf{k}_{12}, \mathbf{k}'') = 1$].

In the optical range the linear dimensions of the system exceed the wavelength of the spontaneous emission. In this case the molecules may be correlated in a way that coherent radiation is observed at different frequencies in directions linked by certain geometric relationships. The Hamiltonian of the interaction of a molecule with the field has the form

$$H_{\text{int}} = -\sum_{\mathbf{k}\lambda} \left(\frac{1}{2\omega V}\right)^{1/2} \{a_{\mathbf{k}\lambda} \mathbf{e}_{\mathbf{k}\lambda} [(\mathbf{e}_1 - i\mathbf{e}_2) I_+(\mathbf{k}) + (\mathbf{e}_3 + i\mathbf{e}_4) K_-(\mathbf{k}) + (\mathbf{e}_5 + i\mathbf{e}_6) L_1(\mathbf{k})]\} + \text{H. conj.}, \tag{11}$$

where $a_{\mathbf{k}\lambda}^+$ and $a_{\mathbf{k}\lambda}$ are the photon creation and annihilation operations. The operators in this equation are given by the equations

$$\begin{aligned} I_\pm(\mathbf{k}') &= \sum_j I_\pm^{(j)} \exp(\pm i\mathbf{k}'\mathbf{r}_j), \\ K_\pm(\mathbf{k}'') &= \sum_j K_\pm^{(j)} \exp(\mp i\mathbf{k}''\mathbf{r}_j), \\ L_\pm(\mathbf{k}''') &= \sum_j L_\pm^{(j)} \exp(\mp i\mathbf{k}'''\mathbf{r}_j) \end{aligned} \tag{12}$$

and, when the condition

$$\mathbf{k}' + \mathbf{k}'' = \mathbf{k}''' \tag{13}$$

holds, they satisfy the permutation relations of an algebra of the SU_3 group. The wave functions of the system of molecules may be chosen to be simultaneously the eigenfunctions of the unperturbed Hamiltonian (2) and of the Casimir operators $F^2(\mathbf{k}', \mathbf{k}'', \mathbf{k}''')$ and $G^3(\mathbf{k}', \mathbf{k}'', \mathbf{k}''')$ made up from Eqs. (12) when condition (13) is satisfied. Here the Casimir operators do not commute with operations of the type (12) which have different wave vector indices from those already chosen. By analogy with the considerations leading to Eq. (10) we find that the resulting wave functions $|s, a, I, I_3, Y\rangle$ with momenta $\mathbf{k}'$, $\mathbf{k}''$, and $\mathbf{k}'''$ are states of the system for which coherent radiation is emitted in directions $\mathbf{k}'$, $\mathbf{k}''$, and $\mathbf{k}'''$ ($\mathbf{k}' + \mathbf{k}'' = \mathbf{k}'''$) at frequencies ω_{12}, ω_{23}, and ω_{13}. The intensities of spontaneous emission in these directions are calculated using Eqs. (10), where J_{12}, $J_{12}^{(0)}$; J_{32}, $J_{32}^{(0)}$; J_{31}, $J_{31}^{(0)}$ must be interpreted as the rates of emission of photons per unit solid angle in the directions of $\mathbf{k}'$, $\mathbf{k}''$, and $\mathbf{k}'''$. Emission in the directions of $\mathbf{k}'$, $\mathbf{k}''$, and $\mathbf{k}'''$ does not change the quantum numbers s and a while emission in other directions causes them to change and causes the coherence relative to $\mathbf{k}'$, $\mathbf{k}''$, and $\mathbf{k}'''$ to be destroyed.

We now calculate the intensities of the coherent responses after pulses of frequencies ω_{12} (wave vector $\mathbf{k}_{12}$) and ω_{13} (wave vector $\mathbf{k}_{13}$) act on an equilibrium system. From Eq. (6) it follows for the mixed states that

$$J_{21}(\mathbf{k}') = J_{21}^{(0)}(\mathbf{k}')\,\mathrm{Tr}\, I_-(\mathbf{k}')\,\rho I_+(\mathbf{k}'),$$
$$J_{32}(\mathbf{k}'') = J_{32}^{(0)}(\mathbf{k}'')\,\mathrm{Tr}\, K_+(\mathbf{k}'')\,\rho K_-(\mathbf{k}''),$$
$$J_{31}(\mathbf{k}''') = J_{31}^{(0)}(\mathbf{k}''')\,\mathrm{Tr}\, L_+(\mathbf{k}''')\,\rho L_-(\mathbf{k}'''),$$

where ρ is the density matrix of the system of molecules after the pulses are applied. The effect of a pulse on a molecule is described by a unitary transformation in the energy space of the molecule. In this case we have

$$\rho(t) = e^{-iH_1 t} T \rho_0 T^{-1} e^{iH_1 t},$$

where

$$T = T_{13} T_{12},$$
$$T_{12} = \exp\{i(\theta_1/2)[\alpha I_-(\mathbf{k}_{12}) + \alpha^* I_+(\mathbf{k}_{12})]\},$$
$$T_{13} = \exp\{i(\theta_2/2)[L_-(\mathbf{k}_{13})\beta^* + L_+(\mathbf{k}_{13})\beta]\} \qquad (|\alpha| = |\beta| = 1).$$

Here the parameters θ_1 and θ_2 are proportional to the area of the first and second pulses and $\sin^2(\theta_1/2)$ and $\sin^2(\theta_2/2)$ are the probabilities of excitation of the molecules by the first and second pulses, respectively, to the first and third level. The calculations yield

$$\left.\begin{aligned} J_{21}(\mathbf{k}') &= J_{21}^{(0)}(\mathbf{k}')\,N\,\{[\sigma_1 \sin^2(\theta_1/2) + \sigma_2 \cos^2(\theta_2/2)] + \\ &\quad + {}^1/_4(\sigma_1 - \sigma_2)^2 \sin^2\theta_1 \cos^2(\theta_2/2)\,[N\Gamma(\mathbf{k}_{12}, \mathbf{k}') - 1]\}, \\ J_{31}(\mathbf{k}''') &= J_{31}^{(0)}(\mathbf{k}''')\,N\,\{\sigma_3 \cos^2(\theta_2/2) + \sin^2(\theta_2/2)\,[\sigma_1 \cos^2(\theta_1/2) + \\ &\quad + \sigma_2 \sin^2(\theta_1/2)] + {}^1/_4 \sin^2\theta_2\,[\sigma_3 - \sigma_1 \cos^2(\theta_1/2) - \\ &\quad - \sigma_2 \sin^2(\theta_1/2)]^2\,[N\Gamma(\mathbf{k}_{13}, \mathbf{k}''') - 1]\}, \\ J_{32}(\mathbf{k}'') &= J_{32}^{(0)}(\mathbf{k}'')\,N\,\{\sigma_3 \cos^2(\theta_2/2) + \sin^2(\theta_2/2)\,[\sigma_1 \cos^2(\theta_1/2) + \sigma_2 \sin^2(\theta_1/2)] + \\ &\quad + {}^1/_4(\sigma_1 - \sigma_2)^2 \sin^2\theta_1 \sin^2(\theta_2/2)\,[N\Gamma(\mathbf{k}_{13} - \mathbf{k}_{12}, \mathbf{k}'') - 1]\}, \end{aligned}\right\} \qquad (14)$$

where $\sigma_i = N_i/N$, $\Gamma(\mathbf{p}_1, \mathbf{p}_2) = |\langle \exp[i(\mathbf{p}_1 - \mathbf{p}_2)\mathbf{r}]\rangle|^2$, and the symbol $\langle\,\rangle$ denotes averaging over the coordinates of the molecules. For a system of large extent $\langle \exp[i(\mathbf{p}_1 - \mathbf{p}_2)\mathbf{r}]\rangle = \delta_{\mathbf{p}_1, \mathbf{p}_2}$. It follows that superradiant pulses will be observed at all frequencies in directions $\mathbf{k}_{13}$ (frequency ω_{12}), $\mathbf{k}_{13} - \mathbf{k}_{12}$ (frequency ω_{23}), and $\mathbf{k}_{13}$ (frequency ω_{13}). The action of one pulse produces an intense response in the same direction. Then the emission is coherent and isotropic at other frequencies.

TABLE 1. Angular distributions of the superradiant responses of a four-level system

Exciting pulses at frequencies						Superradiant responses at frequencies					
ω_{12}	ω_{23}	ω_{13}	ω_{14}	ω_{24}	ω_{34}	ω_{12}	ω_{23}	ω_{13}	ω_{14}	ω_{24}	ω_{34}
12	—	13	14	—	—	12	13—12	13	14	14—12	14—13
12	23	—	14	—	—	12	23	23+12	14	14—12	14—12—23
12	—	—	14	—	34	12	14—34—12	14—34	14	14—12	34
12	23	—	—	—	34	12	23	12+23	12+23+34	34+23	34
12	—	—	—	24	34	12	24—34	12+24—34	24+12	24	34
12	23	—	—	24	—	12	23	12+23	12+24	24	24+23
12	—	13	—	—	34	12	13—12	13	13+34	34+13+12	34
12	—	13	—	24	—	12	13—12	13	12+24	24	24—13+12
—	23	—	14	—	34	14—23—34	23	14—34	14	23+34	34
—	23	—	14	24	—	14—24	23	23+14—24	14	24	24—23
—	23	13	—	—	34	13—23	23	13	13+34	23+34	34
—	23	13	—	24	—	13—23	23	13	13+24+23	24	24—23
—	23	13	14	—	—	13—23	23	13	14	23+14—13	14—13
—	—	13	14	24	—	14—24	13+24—14	13	14	24	14—13
—	—	13	—	24	34	13+34—24	24—34	13	13+34	24	34
—	—	—	14	24	34	14—34	24—34	14	14	24	34

Up to now we have avoided discussion of relaxation processes. Thus, the intensities were independent of time and determined the coherent responses of the system immediately after the exciting pulses. In order to include the time-dependent structure of these responses in the above scheme, it is necessary to include inhomogeneous broadening caused by the difference in the resonance frequencies of the individual radiators due to an inhomogeneity of the fields within the sample (see also [8]). Calculations show that when two pulses with frequencies ω_{12} and ω_{13} act on the medium an echo signal with a frequency of ω_{32} is formed at time $t + (\omega_{13}/\omega_{32})\tau$ in addition to the coherent responses at frequencies ω_{12} and ω_{13} immediately after completion of the pulses at time t and $t + \tau$. The time τ is much less than the irreversible relaxation times of the system. The intensities of these signals are given by Eqs. (14) as before.

An analysis of the case n = 3 yields the basic features of the multilevel problem. The states of a molecular system with n levels are classified according to the irreducible representations $D(p_1, p_2, \ldots, p_{n-1})$ of the SU_n group. The representations $D(N, 0, \ldots, 0)$ correspond to superradiant states. The wave functions of such states are fully symmetric. Superradiant states may be excited by applying pulses to the system. If the wavelengths are shorter than the system dimensions, the intensities of the responses can easily be found with the aid of a generalization of the formalism used above. Then, reducing the n-level problem to subgroups SU_3 of the SU_n group, it is easy to find the angular distribution of the photon echo signals from the geometric relationships among the wave vectors. Table 1 shows the angular distributions of the superradiant responses of a four-level nonequidistant system after pulsed excitation (the indices alone are used instead of the wave vectors; for example, the notation 12 + 14 denotes $\mathbf{k}_{12} + \mathbf{k}_{14}$). The cases in which the angular distributions are independent of the ordering of the pulses are treated in the table.

2. A Method for Calculating Cooperative Scattering Effects

With a generalization of the Dicke theory to the case of multilevel particles it is possible to consider the effect of the collective properties of a system of molecules on processes which are described by higher-order perturbation theory. In the following we shall take into account the collective behavior of the molecules using Raman scattering of light as an example. Usually spontaneous scattering of radiation by isolated particles (atoms or molecules) takes place independently. Such processes have been studied fairly completely in quantum electrodynamics. Meanwhile, physical conditions (for example, a medium excited by coherent fields and not yet returned to its equilibrium state) are possible in which a substantial phase correlation develops among the individual radiators. Then it is necessary to examine the interaction of the electromagnetic field with the entire assembly of particles as a single system when evaluating the scattering.

We shall first examine the problem of calculating the transition probabilities for coherent Raman scattering when the wavelengths of the initial and final photons are large compared to the linear dimensions of the scattering system (of molecules). To do this we shall use the method developed in Section 1, in which the states and operators for n-level molecules are classified with the aid of the SU_n group in the energy space of the system. When cooperative effects are included the molecules behave as a single system characterized by the cooperation numbers introduced above. After establishing a connection with the generalized Dicke technique, which adequately describes cooperative effects in first-order radiative processes, and with previous work [9, 10] on cooperative scattering, we shall consider the case of scattering on pure cooperative states. This discussion will illustrate the nature of the effect and allows us to proceed to the case of mixed states. We shall calculate the probabilities of a single act of scattering on these states (a process of the type $n_L, n_S \rightarrow n_L - 1, n_S + 1$, where n_L, and n_S are the occupation numbers of the laser and scattered photons). For simplicity we shall consider the case of arbitrary three-level molecules, which does not limit the generality of the discussion. (The formulas for the scattering probabilities obtained below remain valid when an arbitrary number of discrete levels and a continuous spectrum are included.) The Hamiltonian of this system has the form

$$H_1 = H + H_{\text{int}}, \tag{15}$$

where H and H_{int} are given by Eqs. (2) and (4).

For the eigenfunctions of the unperturbed Hamiltonian H, we shall use the basis $|s, a, I, I_3, Y\rangle$ of the irreducible representation of the SU_3 group in energy space. Since in a calculation of the Raman scattering probabilities the level "3" is virtual, we immediately find from Eq. (5) that $a = 0$ and $s = N/2 - I$. The probability amplitude for scattering with formation of a Stokes photon has the form

$$A_S = \frac{\langle 2|H'_{\text{int}}|3\rangle\langle 3|H_{\text{int}}|1\rangle}{-\varepsilon/2 - \varepsilon_3 + \omega} + \frac{\langle 2|H_{\text{int}}|3\rangle\langle 3|H'_{\text{int}}|1\rangle}{-\varepsilon/2 - \varepsilon_3 - \omega_S}, \tag{16}$$

where $\langle|H_{\text{int}}|\rangle$ are the matrix elements for absorption of a photon with wave vector **k** and polarization **e**, i.e.,

$$\begin{aligned} \langle 3|H_{\text{int}}|1\rangle &= -i\,(2\pi\omega n_L)^{1/2}\,[\mathbf{e}\,(\mathbf{e}_5 + i\mathbf{e}_6)]\,\langle 3|L_-|1\rangle, \\ \langle 2|H_{\text{int}}|3\rangle &= -i\,(2\pi\omega n_L)^{1/2}\,[\mathbf{e}\,(\mathbf{e}_3 - i\mathbf{e}_4)]\,\langle 2|K_+|3\rangle, \end{aligned} \tag{17}$$

and $\langle|H'_{\text{int}}|\rangle$ are the matrix elements for emission of a Stokes photon ($\mathbf{k}_S$, $\mathbf{e}_S$), i.e.,

$$\begin{aligned} \langle 2|H'_{\text{int}}|3\rangle &= i\,[2\pi\omega_S\,(n_S + 1)]^{1/2}\,[\mathbf{e}_S^*\,(\mathbf{e}_3 - i\mathbf{e}_4)]\,\langle 2|K_+|3\rangle, \\ \langle 3|H'_{\text{int}}|1\rangle &= i\,[2\pi\omega_S\,(n_S + 1)]^{1/2}\,[\mathbf{e}_S^*\,(\mathbf{e}_5 + i\mathbf{e}_6)]\,\langle 3|K_-|1\rangle. \end{aligned} \tag{18}$$

Using Eq. (6) for the probability of Stokes scattering on a system of molecules in state $|I, I_3\rangle$ (we shall omit the indices s and a in what follows) we obtain

$$w_S = w_S^{(0)}\,(I - I_3)\,(I + I_3 + 1), \tag{19}$$

where $w_S^{(0)}$ is the ordinary expression for the probability for a single molecule. For $I_3 = 0$ and $I = I_{max}$, W_S is proportional to the square of the total number of scattering molecules (superscattering). It is easy to see (see below) that for a system of molecules in equilibrium the spontaneous scattering intensity is always proportional to N_1; that is, the scattering is incoherent, as is to be expected. The cooperative effect appears only in a nonstationary system.

Similarly, for the probability of spontaneous scattering of an antistokes photon we have

$$w_{a\text{-}S} = w_{a\text{-}S}^{(0)}\,(I + I_3)\,(I - I_3 + 1). \tag{20}$$

We note that the ratio of the probabilities of Stokes and antistokes processes from the state $|I, I_3\rangle$ is expressed in terms of the quantity $[(I - I_3)(I + I_3 + 1)]/[(I + I_3)(I - I_3 + 1)]$, which for $I = N/2$ gives $[N_1(N_2 + 1)]/[N_2(N_1 + 1)]$. This last result is given in [10].

In general, when the state of the system is a superposition of states with different I and I_3, we have

$$w_S = w_S^{(0)} \sum_{I, I_3} P_{I, I_3}\,(I + I_3)\,(I - I_3 + 1) = w_S^{(0)}\,\langle L_+ K_- K_+ L_-\rangle, \tag{21}$$

where P_{I,I_3} is the probability of finding the system in state $|I, I_3\rangle$ at the moment of scattering.

When the linear dimensions of the system of molecules exceeds the wavelengths of the initial and final photons, which is usually the case under real conditions, the system may be correlated so that for a given wave vector of the scattered photon the scattering is coherent for some particular direction. In this case we have Eq. (16) for the scattering amplitude but with

$$\begin{gathered}\langle 3|L_-(\mathbf{k})|1\rangle, \quad \langle 2|K_-(\mathbf{k})|3\rangle, \\ \langle 2|K_+(\mathbf{k}_S)|3\rangle, \quad \langle 3|L_-(\mathbf{k}_S)|1\rangle\end{gathered} \tag{22}$$

respectively, written in place of the matrix elements on the right sides of Eqs. (17) and (18). Here $\mathbf{k}$ is the wave vector of the incident photon and $\mathbf{k}_S$ is the scattered photon. According to this equation operators from two of the sets in Eq. (12) enter the scattering amplitude and correspond to the following values for the parameters in Eq. (12):

$$\begin{aligned}&1)\ \mathbf{k}'=\mathbf{k}-\mathbf{k}_S, \quad \mathbf{k}''=\mathbf{k}_S, \quad \mathbf{k}'''=\mathbf{k};\\ &2)\ \mathbf{k}'=\mathbf{k}-\mathbf{k}_S, \quad \mathbf{k}''=-\mathbf{k}, \quad \mathbf{k}'''=-\mathbf{k}_S.\end{aligned}$$

The Casimir operators constructed from these two sets of operators commute if only the functions $|s, a; I, I_3, Y\rangle$ with $s = N/2$, $a = 0$, and $Y = N/3$ enter in their region of definition. The wave functions of the system of molecules may be chosen to be simultaneously eigenfunctions of the unperturbed Hamiltonian H and of the Casimir operators constructed with the first and second sets of operators. The states $|s, a, I, I_3, Y\rangle$ with $s = N/2 - I$, $a = 0$, and $Y = N/3$ introduced in this way are states of the system for which scattering in the direction $\mathbf{k}_S$ is coherent. The degree of coherence is specified by a single quantum number I which relates to the chosen wave vectors $\mathbf{k}$ and $\mathbf{k}_S$. Because of Eqs. (16) and (22) evaluation of the scattering amplitudes reduces to computing the matrix elements of the operators

$$\begin{aligned}K_+(\mathbf{k}_S)L_-(\mathbf{k}) &= I_+(\mathbf{k}-\mathbf{k}_S) + \sum_{j\neq j'} K_+^{(j)}L_-^{(j')}e^{i(\mathbf{k}\mathbf{r}_j-\mathbf{k}_S\mathbf{r}_{j'})},\\ K_+(-\mathbf{k})L_-(\mathbf{k}_S) &= I_+(\mathbf{k}-\mathbf{k}_S) + \sum_{j\neq j'} K_+^{(j)}L_-^{(j')}e^{i(\mathbf{k}\mathbf{r}_{j'}-\mathbf{k}_S\mathbf{r}_j)}.\end{aligned}$$

Since the second terms in these formulas make no contribution to the matrix elements when they act on the function $|N/2 - I, 0, I, I_3, N/3\rangle$, the scattering probabilities for such states are calculated according to Eq. (19), where w_S and $w_S^{(0)}$ must be interpreted as scattering rates per unit angle in the $\mathbf{k}_S$ direction when a photon with wave vector $\mathbf{k}$ is incident on the system, i.e.,

$$w_S(\mathbf{k},\ \mathbf{k}_S) = w_S^{(0)}(\mathbf{k},\ \mathbf{k}_S)(I - I_3)(I + I_3 + 1). \tag{23}$$

We now introduce yet another formula which generalizes Eq. (21) to the case of a system of large extent and which follows from Eq. (23):

$$\begin{gathered}w_S(\mathbf{k},\ \mathbf{k}_S) = w_S^{(0)}(\mathbf{k},\ \mathbf{k}_S)\langle 0^+0\rangle,\\ 0 = {}^1/_2[K_+(\mathbf{k}_S)L_-(\mathbf{k}) + K_+(-\mathbf{k})L_-(-\mathbf{k}_S)].\end{gathered} \tag{24}$$

3. Coherent Spontaneous Scattering

A cooperative effect may be caused under real conditions by perturbation of a medium as a result of the propagation of short exciting optical pulses through it. From the following it will be clear that scattering takes place in the usual way for a medium in equilibrium (the scattering probability equals the sum of the scattering probabilities on the individual molecules). If, however, the medium is described by a nonstationary density matrix (with nondiagonal elements in the energy representation) the scattering is coherent. Then it is just the nonstationary character of the density matrix that is important rather than the degree (which may be very small) to which the populations deviate from the equilibrium values.

In computing the scattering probabilities we shall assume that a single-mode laser pulse (plane wave with wave vector $\mathbf{k}$) falls on a medium at equilibrium with temperature T. The intense field acting on the molecule as the pulse passes may be described classically, and the Raman scattering probabilities will be calculated quantum mechanically. For simplicity we shall neglect the change in position of the molecules as the pulse passes. The calculations will be done for the case of three-level molecules, but the final expressions for the Raman scattering probabilities are valid in the general case of arbitrary molecules.

By Eq. (24) the probability of producing a Stokes photon with wave vector $\mathbf{k}_S$ for a mixed state may be written in the form

$$w_S(\mathbf{k},\ \mathbf{k}_S) = w_S^{(0)}(\mathbf{k},\ \mathbf{k}_S)\,\mathrm{Tr}\,O\rho O^+, \tag{25}$$

where ρ is the density matrix of the molecular system at the time of scattering. We shall find the explicit form of the density matrix at a certain time t after the exciting pulse has passed through a given part of the medium (we neglect relaxation processes, assuming $T_2 = \infty$). Due to the passage of the exciting pulse, part of the molecules shift from the first level to the second and the medium loses its equilibrium. The density matrix may be calculated as follows. Let $\psi(0)$ be the wave function of the system before the pulse passes. At time t we have

$$\psi'(t) = e^{-iH_1 t}T\psi(0). \tag{26}$$

Here T is a unitary transformation describing the perturbation of the medium after the pulse passes. The most general form of the operator T may be found if we consider the effect of a pulse on an isolated molecule. The field acting on the j-th molecule is nonzero during the time interval $t_j \le t \le t_j + \tau$, where t_j is the time the pulse arrives at the site of the j-th molecule. Since the wave function may be regarded as a vector in energy space, the unitary transformation describing scattering on a given molecule can be represented as two successive rotations in the L and K subspaces, which, as can easily be seen, reduce to a rotation in the I-spin subspace. An arbitrary rotation in this latter space is the product of two rotations: (1) about an axis perpendicular to the quantitation axis for I_3 and (2) about axis 3. The operator T, therefore, has the form of a product,

$$T = \exp\left(i\sum_j H_1^{(j)}t_j\right)\prod_{s=1}^{N}\exp\left[\frac{i}{2}\varphi_s(\alpha_s I_+^{(s)} + \alpha_s^* I_-^{(s)}) + i\varphi_s' I_3^{(s)}\right]\exp\left(-i\sum_j H_1^{(j)}t_j\right). \tag{27}$$

Here φ_s and φ_s' are the parameters of the first and second rotations, respectively, in the space of the s-th molecule and depend on τ. The phase α_s determines the location of the axis about which the first rotation takes place ($|\alpha_s| = 1$). The parameters φ_s' disappear in the later calculations; hence they will be assumed equal to zero in the following. The quantity $\sin^2(\varphi_s/2)$ is the probability of a molecular transition from the first level to the second due to the passage of the exciting pulse. In general the quantities φ_s are not the same for different molecules. After substituting $t_j = \mathbf{kr}_j/\omega$ in Eq. (27) we obtain

$$T = \exp\left\{i\sum_j\left[\alpha_j\frac{1}{2}\varphi_j I_+^{(j)}\exp\left[i(\omega_{12}/\omega)\,\mathbf{kr}_j\right] + \alpha_j^*\frac{1}{2}\varphi_j I_-^{(j)}\exp\left[-i(\omega_{12}/\omega)\,\mathbf{kr}_j\right]\right]\right\} \qquad (\omega_{12} = \varepsilon_2 - \varepsilon_1).$$

Transforming from Eq. (26) to a density matrix description of the scattering we find

$$\rho(t) = e^{-iH_1 t}T\rho_0 T^{-1}e^{iH_1 t}, \tag{28}$$

where ρ_0 is the equilibrium density matrix, given by

$$\rho_0 = \frac{\exp(-H_1/kT)}{\operatorname{Tr}\exp(-H_1/kT)} = \prod_j (\sigma_1 P_1^{(j)} + \sigma_2 P_2^{(j)}),$$
$$\sigma_i = \bar{N}_i/N = {}^1/_2[1 + (-1)^{i-1}\tanh(\varepsilon/2kT)] \quad (i = 1,\ 2),$$
$$P_1^{(j)} = 2L_3^{(j)} + (1/3 - Y)^{(j)}, \qquad P_2^{(j)} = 2K_3^{(j)} + (1/3 - Y)^{(j)}.$$

In the following $T\rho_0 T^{-1}$ may be conveniently represented in the form $\prod_j A_j$, where

$$A_j = [\sigma_1 \cos^2(\varphi_j/2) + \sigma_2 \sin^2(\varphi_j/2)]\, P_1^{(j)} + [\sigma_1 \sin^2(\varphi_j/2) + \sigma_2 \cos^2(\varphi_j/2)]\, P_2^{(j)} + {}+ {}^1/_2 i(\sigma_1 - \sigma_2)\sin\varphi_j \cdot [\alpha_j I_+^{(j)} e^{i(\omega_{12}/\omega)\mathbf{k}\mathbf{r}_j} - \alpha_j^* I_-^{(j)} e^{-i(\omega_{12}/\omega)\mathbf{k}\mathbf{r}_j}]. \tag{29}$$

This equation gives an expression for the density matrix of the medium after it is perturbed by the exciting pulse. The nondiagonal elements in Eq. (29) describe cooperative scattering.

As an illustration we shall now calculate one of the four terms in Eq. (25), for example,

$$u = \operatorname{Tr}\rho L_+(\mathbf{k})\, K_-(\mathbf{k}_S)\, K_+(\mathbf{k}_S)\, L_-(\mathbf{k}) = \sum_{j_1 j_2 l_1 l_2} \operatorname{Tr}\prod_j A_j L_+^{(j_1)} K_-^{(j_2)} K_+^{(l_1)} L_-^{(l_2)} e^{i\mathbf{k}(\mathbf{r}_{l_2} - \mathbf{r}_{j_1})} e^{i\mathbf{k}_S(\mathbf{r}_{j_2} - \mathbf{r}_{l_1})}. \tag{30}$$

With the aid of Eq. (29) it is easy to see that only the terms with $j_1 = j_2 = l_1 = l_2$ and $j_1 = j_2 \neq l_1 = l_2$ are nonzero in Eq. (30). Assuming that φ is a slowly varying function of the coordinates, we obtain

$$u = N[\sigma_1 \langle\cos^2(\varphi/2)\rangle + \sigma_2 \langle\sin^2(\varphi/2)\rangle] + N(\sigma_1 - \sigma_2)^2\ [N\langle\sin\varphi\rangle^2 \delta_{\mathbf{k}_S,\,(\omega_S/\omega)\mathbf{k}} - \langle\sin^2\varphi\rangle].$$

Here, since $\lambda \ll a$, it was assumed that $\langle\exp\{i[\mathbf{k}_S(\omega_S/\omega)\mathbf{k}]\,\mathbf{r}\}\rangle = \delta_{\mathbf{k}_S,\,(\omega_S/\omega)\mathbf{k}}$. The quantity $\langle\sin\varphi\rangle^2 = \left(\frac{1}{N}\sum_j \sin\varphi_j\right)^2$ is an integral characteristic of the medium which determines how much it is perturbed. The remaining terms of Eq. (25) are calculated similarly. Finally, for the probability of Stokes scattering we obtain

$$w_S(\mathbf{k},\ \mathbf{k}_S) = w_S^{(0)}(\mathbf{k},\ \mathbf{k}_S)\,\{N[\sigma_1 \cos^2(\varphi/2) + \sigma_2 \sin^2(\varphi/2)] + N\sin^2\varphi \cdot (\sigma_1 - \sigma_2)^2\,[N\delta_{\mathbf{k}_S,\,(\omega_S/\omega)\mathbf{k}} - 1]\}. \tag{31}$$

For $\varphi = 0$ (an equilibrium medium) Eq. (31) yields the usual result, $w_S(\mathbf{k}, \mathbf{k}_S) = w_S^{(0)}(\mathbf{k}, \mathbf{k}_S) N\sigma_1$. In the same way expressions are found for the probability of antistokes scattering:

$$w_{a-S}(\mathbf{k},\ \mathbf{k}_{a-S}) = w_{a-S}^{(0)}(\mathbf{k},\ \mathbf{k}_{a-S})\,\{N(\sigma_1 \sin^2(\varphi/2) + \sigma_2 \cos^2(\varphi/2)] + N\sin^2\varphi \cdot (\sigma_1 - \sigma_2)^2\,[N\delta_{\mathbf{k}_{a-S},\,(\omega_{a-S}/\omega)\mathbf{k}} - 1]\}. \tag{32}$$

This discussion shows that spontaneous Raman scattering of light on a nonstationary medium may be coherent. Such scattering occurs when the medium is initially excited by coherent fields and occupies an intermediate position between ordinary spontaneous scattering and stimulated Raman scattering. We note that an analysis based on the combined use of the equations of motion for the density matrix and Maxwell's equations [17] also yields Eqs. (31) and (32). It is shown in [17] that coherent spontaneous Raman scattering may be used in active Raman scattering spectroscopy when the probing pulse is scattered by a medium initially (or simultaneously) excited by coherent fields. Resonance or scattering pulses may be used for the coherent fields. Under these conditions the probe pulse is scattered coherently and this may be an effective method for obtaining spectroscopic information [18].

4. Superradiance Effects in Equidistant Systems

In the preceding sections we have considered a system of molecules with a nonequidistant spectrum. The group theoretical approach is also very effective in the case of equidistant radiators. We now discuss superradiance effects in a system of oscillators.

It is known that the bilinear combinations of boson creation and annihilation operators, $L_+ = {}^1/_2(a^+)^2$, $L_- = {}^1/_2 a^2$, and $L_3 = (a^+a + 1/2)/2$, form a Lie algebra in the group SU(1,1). The basis $|n\rangle = \frac{(a^+)^n}{\sqrt{n!}}|0\rangle$ in the space of states of each oscillator may be regarded as the basis of a unitary positive discrete series in SU (1,1): Then $\mathcal{F} = {}^1/_4\ ({}^3/_4)$ if n is even (odd) and $\mathfrak{M} = {}^1/_2(n + {}^1/_2)$ takes the values $\mathcal{F}, \mathcal{F}+1, \ldots$. For a system of N oscillators with an unperturbed Hamiltonian

$$\mathcal{H} = 2\hbar\omega_0 \sum_{j=1}^{N} L_3^{(j)} \tag{33}$$

the union of the generators of SU(1.1), the group of isolated oscillators, reduces to the operators $L_\alpha = \sum L_\alpha^{(j)}$ which also form an algebra in SU (1,1): $[L_+, L_-] = 2L_3$, $[L_3, L_\pm] = \pm L_\pm$. In view of Eq. (33) each energy state of the system may be classified according to the SU (1,1) group. The transformation from the basis $|\mathcal{F}_1\mathfrak{M}_1\rangle\, |\mathcal{F}_2\mathfrak{M}_2\rangle \ldots |\mathcal{F}_N\mathfrak{M}_N\rangle$ to irreducible representations of the group is made in a way similar to the technique for the angular momenta of the SU(2) group:

$$|(\mathcal{F}_1\mathcal{F}_2\ldots\mathcal{F}_N)^s\mathcal{F}_{ij};\ L,\ L_3\rangle =$$

$$= \sum_{\mathfrak{M}_1\ldots\mathfrak{M}_N} \left(\begin{matrix}\mathcal{F}_1\mathcal{F}_2\ldots\mathcal{F}_N\\ \mathfrak{M}_1\mathfrak{M}_2\ldots\mathfrak{M}_N\end{matrix}\middle|(\mathcal{F}_1\mathcal{F}_2\ldots\mathcal{F}_N)^s\mathcal{F}_{ij};\ L,\ L_3\right)|\mathcal{F}_1\mathfrak{M}_1\rangle\ldots|\mathcal{F}_N\mathfrak{M}_N\rangle,$$

where $\left(\begin{matrix}\mathcal{F}_1\ldots\mathcal{F}_N\\ \mathfrak{M}_1\ldots\mathfrak{M}_N\end{matrix}\middle|(\mathcal{F}_1\ldots\mathcal{F}_N)^s\mathcal{F}_{ij};\ L,\ L_3\right)$ is the generalized Clebsch–Gordan coefficient of the SU(1,1) group, s is the coupling mode, $\mathcal{F}_{ij}$ are the intermediate momenta, and $L(L-1)$ is the eigenvalue of the Casimir operator, $L_3 = \sum_j \mathfrak{M}_j$

For a given energy $2\hbar\omega_0L_3$ the cooperative number L takes the values $L_3, L_3-1, \ldots, \mathcal{F}_1+\mathcal{F}_2+\ldots+\mathcal{F}_N$. In the dipole approximation the interaction with the field has the form

$$V^{(1)} = -\frac{ieA}{c}\sqrt{\frac{N\hbar\omega}{2m}}(A+A^+), \quad A = \frac{1}{\sqrt{N}}\sum_j a_j, \quad [A,\ A^+] = 1. \tag{34}$$

The pair of operators A^+ and A is an irreducible tensor of rank $^1/_2$ relative to the SU (1,1) group, and $A^+(A)$ corresponds to the component $^1/_2(-^1/_2)$; hence,

$$A^+|(\mathcal{F}_1\ldots\mathcal{F}_N)^s\mathcal{F}_{ij};\ L,\ L_3\rangle = \sum_{L'}\left(\tfrac{1}{2}L\,\tfrac{1}{2}L_3|L'L_3'+\tfrac{1}{2}\right)(L'\|A\|L)\,|(\mathcal{F}_1'\ldots\mathcal{F}_N')^{s'}\mathcal{F}_{ij}';\ L',\ L_3'+\tfrac{1}{2}\rangle \tag{35}$$

(the formula is similar for the operator A). The Clebsch–Gordan coefficient in Eq. (35) relates the unitary discrete series $|L, L_3\rangle$ to the nonunitary finite-dimensional representation $|{}^1/_2 Q\rangle$. The moduli of the reduced matrix elements are unity. When $L = L_3$ the operator A has only one nonzero matrix element. Then the intensity of the radiation is $I = \gamma L_3 N = \gamma\varepsilon N^2$ (where $\gamma = 2\hbar e^2\omega_0^3/3mc^3$ and $\varepsilon = L_3/N$ is the mean energy of an oscillator), or N times larger than the emitted intensity from an incoherent state. For example, for a thermodynamically equilibrium state we have Eq. (41) (see below), where $\alpha_l = 0$. In superradiant states with $L = L_3$ the average values of the operators $\Delta x^2 = \sum_{i<j}\frac{(x_i - x_j)^2}{N(N-1)}$ and $\Delta p^2 = \sum_{i<j}\frac{(p_i - p_j)^2}{N(N-1)}$, where x_j, p_j are the coordinate and momentum of the j-th operator, have the minimum possible values equal to $\hbar/2m\omega_0$ and $m\hbar\omega_0/2$. This corresponds to the classical picture of superradiance [8] (when the oscillators move with identical amplitudes and phases the system is equivalent to a single oscillator with charge Ne and $I = \gamma\varepsilon N^2$).

Although this classification of a system of oscillators makes it possible to establish a direct relationship between single-particle and cooperative states and to use well-developed angular momentum techniques, It is unsuitable in two respects: (1) for a complete classification of the basis it is necessary to specify $\mathcal{F}_1, \mathcal{F}_2, \ldots, \mathcal{F}_N, \mathcal{F}_{ij}$ along with L and L_3, and (2) the operators A and A^+ generally have two nonzero matrix elements. We now introduce the operators

$$K_+ = \frac{1}{2} A^+A^+, \quad K_- = \frac{1}{2} AA, \quad K_3 = \frac{1}{2}\left(A^+A + \frac{1}{2}\right),$$

which obey the commutation relations of the SU(1,1) algebra. Since, in addition, $[K_3, \mathscr{H}]=0$ and $[\mathscr{H}, K_\pm] = \pm 2\hbar\omega_0 K_\pm$, the operators $K_\pm$, K_3, and $\mathscr{H}$ form a Lie algebra of the group of second rank and of fourth order. The operators $P_3 \equiv \frac{\mathscr{H}}{2\hbar\omega_0} - K_3$, $P_\pm \equiv L_\pm - K_\pm$, and $\mathscr{H}$ form just such an algebra with $[P_\alpha, K_\alpha] = 0$. The classifications of the system of oscillators in terms of the K_α, $\mathscr{H}$ and P_α, $\mathscr{H}$ groups are practically the same since it is possible to use the operators $I_1 = K_3^2 - {}^1/_2 (K_+K_- + K_-K_+)$ and P_3 for the two Casimir operators of the first group and the operators $I_2 = P_3^2 - {}^1/_2(P_+P_- + P_-P_+)$ and K_3 for the Casimir operators of the second. For concreteness we shall use the basis $|I_1P_3; R, M\rangle$ of the first group, where R and M are the eigenvalues of the operators K_3 and $\mathscr{H}$ (in what follows the values of the Casimir operators are omitted), i.e.,

$$\mathscr{H}\,|R, M\rangle = \hbar\omega_0\left(M + \frac{N}{2}\right)\Big|R, M\rangle \quad \left(M = \sum_j n_j\right),$$

$$K_3\,|R, M\rangle = \frac{1}{2}\left(R + \frac{1}{2}\right)\Big|R, M\rangle, \quad R = 0, 1, \ldots, M.$$

To construct the basis $|R, M\rangle$ we act on the vacuum state $|00\rangle$ with the operators A^+, P_+, and K_+:

$$A^+\,|R, M\rangle = \sqrt{R+1}\,|R+1, M+1\rangle,$$

$$2K_+\,|R, M\rangle = \sqrt{(R+1)(R+2)}\,|R+2, M+2\rangle,$$

$$2P_+\,|R, M\rangle = \begin{cases} [M(M+2) - 2R(M+2) + (M+2)(N-1)]^{1/2}\,|R, M+2\rangle & (M \text{ — even}), \\ [(M+1)^2 - 2R(M+1) + (M+1)(N-1)]^{1/2}\,|R, M+2\rangle & (M \text{ — odd}). \end{cases}$$

Here we obtain states in which R and M have the same parity. They are symmetric relative to permutation of the oscillators; for example, for M = R,

$$|R, M\rangle_{R=M} = \sum_{\Sigma n_j = M} \left(\frac{M!}{n_1!\, n_2! \ldots n_N!}\right)^{1/2} |n_1, n_2 \ldots, n_N\rangle. \qquad (36)$$

States with different parities of R and M may be obtained using the operators $c_k^+ = [k(k+1)]^{-1} \times \left[\sum_{j=1}^{k} a_j^+ - k a_{k+1}^+\right]$ and $[c_k, c_{k'}^+] = \delta_{k,k'}$ given in [19]. The intensity of emission from the state $|R, M\rangle$ is $I = \gamma RN$. From this it follows that the states of Eq. (36) are superradiant.

Classifying the system according to the groups K_α, $\mathscr{H}$ and P_α, $\mathscr{H}$ is suitable not only because of the simplicity of the matrix elements for emission in the dipole approximation, but because of the fact that the operators A and A^+ commute with the operator P_3. Its eigenvalue, equal to ${}^1/_2(M - R) + {}^1/_4(N - 1)$, does not change during the interaction with the electromagnetic field. Thus, for example, it is clear that the effect of a resonant pulse on the ground state, for which M = R = 0, is to transform the system to a superradiant state (36). Because of the existence of an integral of the motion the procedure for evaluating the induced and echo signals in a system with a bounded equidistant spectrum is greatly simplified. Without dwelling on this case we now calculate the coherent responses of a system of oscillators after a pulsed interaction with externally applied electromagnetic pulses. We note that a number of important results for similar systems (Landau levels) were obtained in [20].

Let L coherent pulses of frequency ω_0 with wave vectors $\mathbf{k}_1$ (at time τ_1), $\mathbf{k}_2(\tau_2)$, ..., $\mathbf{k}_L(\tau_L)$ act successively on an equilibrium system. The radiant intensity at time t per unit solid angle in direction $\mathbf{k}$ is equal to

$$I(\mathbf{k}, t) = I_0(\mathbf{k}) \operatorname{Tr} \rho(t) \sum_{i,j} f_i^+ f_j, \tag{37}$$

where for a dipole interaction between the pulses and the material, $f_i = a_i \exp(-i\mathbf{k}\mathbf{r}_i)$. In the following we shall use the Glauber state representation, in which the equilibrium density matrix has the form

$$\rho_0 = \int |\{\beta\}\rangle\langle\{\beta\}| \prod_{k=1}^{N} (\pi \langle n\rangle)^{-1} \exp\left(-\frac{|\beta_k|^2}{\langle n\rangle}\right) d^2\beta_k,$$

$$\langle n\rangle = \left[\exp\left(\frac{\hbar\omega_0}{kT}\right) - 1\right]^{-1}, \qquad |\{\beta\}\rangle = \prod_{k=1}^{N} |\beta_k\rangle_k,$$

$$|\beta_k\rangle = \exp\left(-\frac{|\beta|^2}{2}\right) \sum_{p=0}^{\infty} \frac{\beta^p}{\sqrt{p!}} |p\rangle \left(a^+ a |p\rangle = p|p\rangle\right).$$

The Hamiltonian $\mathcal{H} + V^{(1)}$ describes a system of oscillators in the presence of an external force. Its effect amounts to a displacement of the oscillator by the displacement operator

$$D_1(\alpha) = \prod_{j=1}^{N} D_1^{(j)}(\alpha), \qquad D_1^{(j)}(\alpha) = \exp(\alpha a_j^+ - \alpha^* a_j)$$

($|\alpha|^2 = \left|\frac{eA\omega}{c}\right|^2 \frac{1}{2m}$ is the average excitation energy of the oscillator). From this we find that immediately after application of the l-th pulse

$$\rho_l = T_l \rho_l' T_l^{-1},$$

$$T_l = \exp\left(\frac{i}{\hbar} \sum_j \mathcal{H}^{(j)} t_j^l\right) \prod_j D_1^{(j)}(\alpha_l) \exp\left(-\frac{i}{\hbar} \sum_j \mathcal{H}^{(j)} t_j^l\right), \tag{38}$$

$$\mathcal{H}_j = \hbar\omega_0 \left(a_j^+ a_j + \frac{1}{2}\right),$$

where $t_j^l = \mathbf{k}_l \mathbf{r}/\omega_0$ is the time for the l-th pulse to reach the location of the j-th molecule and ρ_l' is the density matrix immediately before passage of the l-th pulse. The evolution of the system under the influence of the internal perturbations which determine the reversible transverse relaxation is described by the Hamiltonian $\mathcal{H}^* = \sum_j \mathcal{H}_j^*$; $\mathcal{H}_j^* = \hbar(\omega_0 - \omega_j)\left(a_j^+ a_j + \frac{1}{2}\right)$. Thus,

$$\rho_{l+1}' = T_{l+1,l} \rho_l T_{l+1,l}^{-1}, \qquad T_{l+1,l} = \exp\left[-\frac{i}{\hbar} \sum_j \mathcal{H}_j^* (\tau_{l+1} - \tau_l)\right]. \tag{39}$$

Using the equations

$$\exp(i\delta a^+ a) D_1(\alpha) = D_1(\alpha e^{i\delta}) \text{ and } D_1(\alpha) D_1(\beta) = \exp\left(\frac{\alpha\beta^* - \alpha^*\beta}{2}\right) D_1(\alpha + \beta),$$

from Eqs. (38) and (39) we obtain

$$\rho_{l+1}' = J_{l+1,l} \rho_l' J_{l+1,l}^{-1} \text{ and } J_{l+1,l} = D_1(\alpha_1 \exp[i\mathbf{k}_l \mathbf{r}_j - i(\omega_j - \omega_0)(\tau_{l+1} - \tau_l)]).$$

At time t ($t > \tau_L$)

$$\rho(t) = J_{t,0} \rho_0 J_{t,0}^{-1} \text{ and } J_{t,0} = \prod_j D_1^{(j)} \left\{\sum_l \alpha_l \exp[i\mathbf{k}_l \mathbf{r}_j - i(\omega_j - \omega_0)(t - \tau_l)]\right\}. \tag{40}$$

Substituting Eq. (40) in (37) and using the equations $|\beta\rangle = D_1(\beta)|0\rangle$, $a|\beta\rangle = \beta|\beta\rangle$, $\int (\alpha^* + \beta^*)^n \beta^m \times \exp(-|\beta|^2)\, d^2\beta = \pi (\alpha^*)^n + n\pi\delta_{nm}$, we find that for $t = \tau_l$ the induced signal is

$$I = \gamma N \left[\langle n \rangle + \sum_{l=1}^{l} |\alpha_l|^2 \right] + \gamma |\alpha_{l_1}|^2 \{ N^2 | \langle \exp[i(\mathbf{k}_l - \mathbf{k})\mathbf{r}] \rangle |^2 - N \}. \tag{41}$$

Here there are no echo signals as opposed to the case when there is a quadrupole interaction of the pulses with the system of oscillators. An oscillator acted on by a quadrupole force is described by the Hamiltonian

$$\hbar\omega_0 \left(a^+ a + \frac{1}{2} \right) - \frac{F\hbar}{4m\omega_0} (a + a^+)^2, \qquad F = \frac{2\,|\nabla E|}{9},$$

which transforms to the diagonal form $\hbar\omega'(b^+ b + 1/2)$, $\omega' = (\omega_0^2 - F/m)^{1/2}$ with the aid of the canonical transformation

$$b = ua + \sqrt{u^2 - 1}\, a^+, \qquad b^+ = ua^+ + \sqrt{u^2 - 1}\, a, \qquad [b, b^+] = 1,$$

$$1 - 2u^2 = \frac{\varkappa}{\sqrt{\varkappa^2 - 4}}, \qquad \varkappa = 2 - \frac{4m\omega_0^2}{F}.$$

From this it follows that the ground state $|0\rangle'$ of an unperturbed oscillator with frequency ω' satisfies the equation $a|0\rangle' = \mu a^+ |0\rangle'$, where $\mu = -\sqrt{(u^2 - 1)/u^2}$. The solution of this equation has the form

$$|\mu\rangle = \frac{(1 - |\mu|^2)^{1/4}}{\sqrt{\Gamma(1/2)}} \sum_{p=0}^{\infty} \mu^p \sqrt{\frac{\Gamma(p + 1/2)}{p!}}\, |2p\rangle = (1 - |\mu|^2)^{1/4} \exp\left(\frac{\mu a^{+2}}{2} \right) |0\rangle \equiv D_2(\mu)|0\rangle. \tag{42}$$

It can be shown that the action of the pulse transforms the vacuum state (assuming $kT \ll \hbar\omega_0$) into state (42) [where $(|\mu|^2 = E/(E + 1)$ and $\hbar\omega_0 E$ is the energy transferred by the pulse to the oscillator]. The intensity of the quadrupole radiation from the system is given by Eq. (37) with $f_i = a_i^2 \exp(-i\mathbf{k}\mathbf{r}_i)$ (no dipole radiation is present). Using the properties of the operator $D_2(\alpha) = \exp[(\alpha a^{+2} - \alpha^* a^2)/2]$ it is possible to show, as with Eqs. (38)-(40), that the wave function of the system at time $t > \tau_L$ is

$$\prod_j |\nu^j(t)\rangle, \quad \nu^j(t) = \frac{1}{2} \sum_{l=1}^{L} \prod_{p_1 < p_2 < \ldots < p_{\psi(l)}}^{l} \nu_{p_1}^j \nu_{p_2}^{j*} \nu_{p_3}^j \ldots \nu_{p_{\psi(l)}}^j \left(1 + \frac{1}{2} \sum_{l=2}^{L} \prod_{p_1 < p_2 < \ldots < p_{\varphi(l)}}^{l} \nu_{p_1}^j \nu_{p_2}^{j*} \nu_{p_3}^j \nu_{p_4}^{j*} \ldots \nu_{p_{\varphi(l)}}^{j*} \right)^{-1},$$

$$\psi(l) = 2\left(l - \left[\frac{l}{2} \right] \right) - 1, \qquad \varphi(l) = 2\left(l - \left[\frac{l+1}{2} \right] \right),$$

$$\nu_l^j = \mu_l \exp\left\{ i2\mathbf{k}_l \mathbf{r}_j \frac{\omega_0}{\omega_l} - 2i(\omega_j - \omega_0)(t - \tau_l) \right\}. \tag{43}$$

where ω_l is the frequency of the l-th pulse. The matrix element of the operator a_j^2 is equal to $\langle \mu_j | a_j^2 | \mu_j \rangle = \frac{\mu_j}{1 - |\mu_j|^2}$. After substituting Eq. (43) into (37) and retaining terms $\sim |\mu|^6$ ($|\mu_l| < 1$) in the radiant intensity, we find that, besides the induced signals at time $t = \tau_l$ with wave vectors $\mathbf{k} = 2\mathbf{k}_l(\omega_0/\omega_l)$, the system generates echo signals. The times they occur and their wave vectors are

$$t_n = \tau_l + \tau_{l_2} - \tau_{l_1}, \qquad \mathbf{k} = 2\left(\mathbf{k}_l \frac{\omega_0}{\omega_l} + \mathbf{k}_{l_2} \frac{\omega_0}{\omega_{l_2}} - \mathbf{k}_{l_1} \frac{\omega_0}{\omega_{l_1}} \right), \tag{44}$$

$$l,\ l_1,\ l_2 = 1, 2, \ldots, L.$$

In using Eq. (44) it is necessary that the indices l, l_1, and l_2 take only those values for which t_n successively becomes greater than the values $\tau_1, \tau_2, \ldots, \tau_L$. The overall number of the most intense echo signals is $L(L^2 - 1)/6$.

LITERATURE CITED

1. I. D. Abella, N. A. Kurnit, and S. R. Hartmann, Phys. Rev., 141:391 (1966).
2. U. Kh. Kopvillem and V. R. Nagibarov, Fiz. Met. Metalloved., 15:313 (1963).
3. S. L. McCall and E. L. Hahn, Phys. Rev., 183:457 (1969).
4. R. H. Dicke, Phys. Rev., 93:99 (1954).
5. G. S. Kruglik and A. I. Apanasevich, Izv. Akad. Nauk SSSR, Ser. Fiz., 27:483 (1963).
6. L. A. Shelepin, Zh. Éksp. Teor. Fiz., 54:1963 (1968).
7. É. Sh. Teplitskii, Teor. Mat. Fiz., 2:399 (1970).
8. V. R. Nagibarov and N. K. Solovarov, Fiz. Tverd. Tela, 11:1136 (1969); Phys. Status Solidi, 37:889 (1970).
9. G. S. Kruglik, Opt. Spektrosk., 19:171 (1965).
10. D. F. Walls, Z. Phys., 224:117 (1971).
11. T. M. Makhviladze and L. A. Shelepin, FIAN Preprint No. 34 (1972); Phys. Lett., 39A:409 (1972).
12. T. M. Makhviladze and L. A. Shelepin, FIAN Preprint No. 145 (1971); Zh. Éksp. Teor. Fiz., 62:2066 (1972).
13. I. M. Lizin, T. M. Makhviladze, and L. A. Shelepin, Izv. Akad. Nauk SSSR, Ser. Fiz., 37:2224 (1973).
14. T. M. Makhviladze and L. A. Shelepin, Phys. Rev., 9A:538 (1974).
15. T. M. Makhviladze and L. A. Shelepin, Trudy FIAN, 70:120 (1973).
16. J. de Swart, Usp. Fiz. Nauk, 84:533 (1964).
17. T. M. Makhviladze, M. E. Sarychev, and L. A. Shelepin, FIAN Preprint No. 18 (1975); Zh. Éksp. Teor. Fiz., 69:499 (1975).
18. S. A. Akhmanov and N. I. Koroteev, Zh. Éksp. Teor. Fiz., 67:1306 (1974).
19. J. Katriel and G. Adam, Phys. Lett., 33A:190 (1970); W. Zacowz, Phys. Lett., 32A:87 (1970).
20. Ya. Ya. Asadullin and U. Kh. Kopvillem, Fiz. Tverd. Tela, 9:2737 (1967); Zh. Éksp. Teor. Fiz., 61:2487 (1971).

COHERENCE AND RADIATION CONTROL IN MULTILEVEL SYSTEMS

V. P. Kudrya, T. M. Makhviladze, I. G. Sinitsyn, and L. A. Shelepin

The problem of controlling laser systems by radiation when multilevel molecules interact coherently with a field is examined. Equations describing the kinetics of coherent relaxation are derived. The examples considered include radiation in a cavity and in free space, and the propagation and interaction of ultrashort pulses in a resonant medium. Other applications of the theory developed here are considered briefly.

INTRODUCTION

The use of lasers is steadily increasing. Lasers are used in technology and for information transmission. In both cases there is a need to tune, modulate, and control the radiation.

Usually the operating regime of a laser is determined by the specific parameters of the device. To change to another regime requires either a substantial change in these parameters or even a new laser. This is typical of the usual two-level system, where only the inverted population can be varied in order to control the gain or output.

The situation changes if we consider a real multilevel system under coherent interaction conditions when the characteristic interaction time of the molecules with the field does not exceed the polarization damping time due to relaxation processes. Here it is possible to act with signals at other frequencies on the processes causing lasing or on the light propagating in the medium.

It should be emphasized that control is achieved in this case mainly by phase correlation of the radiating molecules due to coupling of the molecules through the light field, rather than by changes in the populations.

Correlation of radiation from various transitions under coherent interaction conditions has been discussed previously [1, 2].

In this article we develop a method for analyzing a system of multilevel radiators which permits examination of the interaction of radiation at various frequencies and allows us to study the controlling effect of various pulses. A number of features of multilevel systems will be discussed using the example of a three-level system. As will become obvious later, in a three-level system the role of the intermediate level, in the usual sense, may be compared to the role of the grid in a triode.

Two basic approaches to the task of building controlled lasers may be distinguished. They are (1) control by acting on the active medium inside the laser, and (2) control by the action of auxiliary pulses on a multilevel medium located outside the laser as the main laser pulse passes through this medium (in particular, under self-transmission conditions). Below, in Sections 1-3, we consider questions associated with the first approach; in Section 4, with the second. In Section 5 we discuss briefly some other applications of the theory in which there may be significant coherent effects.

1. Multilevel Molecules in a Field.

Basic Equations

To analyze the coherent effects produced by the interaction of a system of multiatomic molecules with an electromagnetic field we shall use the group theoretical approach based on a description of multilevel nonequidistant systems with the SU_n group [3, 4].

We first introduce a number of results from the quantum theory of relaxation [5]. Let the Hamiltonian of the system being studied have the form E + F + V, where E and F are the Hamiltonians of the dynamic and dissipative subsystems and $V=\sum_{i=1}^{N} v_i w_i$ is the Hamiltonian of their interaction (v_i and w_i are operators acting on the variables of the dynamic and dissipative subsystems, respectively, and N is the total number of particles in the dynamic subsystems). After Fourier expansion of v_i with the eigenfrequencies of the dynamic subsystem,

$$v_i = \sum_r v_i^r e^{i\omega_r t},$$

we obtain the equations of motion for the average value of an arbitrary operator O of the dynamic subsystem (neglecting level shifting):

$$\langle \dot{O} \rangle = i\langle [E, O] \rangle + \sum_{r,s,i,i'} \Phi_{ii'}^{rs} \langle v_{i'}^s [O, v_i^r] + [v_{i'}^s, O] v_i^r \rangle, \tag{1}$$

where

$$\Phi_{ii'}^{rs} = \pi \sum_{\alpha\alpha'} \delta(\omega_s + \omega_{\alpha\alpha'}) \langle \alpha | w_i^r | \alpha' \rangle \langle \alpha' | w_{i'}^s | \alpha \rangle P_\alpha, \quad \hbar = 1,$$

α enumerates the states of the dissipative subsystem, $\omega_{\alpha\alpha'} = F_\alpha - F_{\alpha'}$, and P_α is the diagonal part of the density matrix of the dissipative subsystem. We shall use Eq. (1) to analyze a system of n-level molecules with a nonequidistant spectrum which interacts with resonant electromagnetic fields.

The Hamiltonian of such a system can be represented [3, 4] in the form of a linear combination of generators $A_{\mu\nu}$ of the SU_n group which act in the space of the molecular subsystem:

$$H = \sum_{\mu,\nu}^{n} h_{\mu\nu} A_{\mu\nu}, \quad A_{\mu\nu} = \sum_{i=1}^{N} A_{\mu\nu}^{(i)}, \quad h_{\mu\nu} = \sum_{i=1}^{N} h_{\mu\nu}^{(i)}, \quad h_{\mu\nu}^{*} = h_{\nu\mu},$$

$$[A_{\mu\nu}, A_{\sigma\rho}] = \delta_{\mu\rho} A_{\sigma\nu} - \delta_{\sigma\nu} A_{\mu\rho}, \quad \sum_\mu A_{\mu\mu} = 0, \quad A_{\mu\nu}^{+} = A_{\nu\mu}.$$

In the dipole approximation $h_{\mu\nu}$ are expressed in terms of the dipole moment and vector potential operators at the system location.

Regarding the field as a dissipative subsystem we have

$$E = \sum_\nu h_{\nu\nu} A_{\nu\nu}, \quad V = \sum_{\mu \neq \nu} h_{\mu\nu} A_{\mu\nu}.$$

In this case Eq. (1) takes the form

$$\langle \dot{A}_{\sigma\rho} \rangle = i(h_{\rho\rho} - h_{\sigma\sigma}) \langle A_{\sigma\rho} \rangle + \sum_{\mu \neq \sigma} (\Phi^{\mu\sigma\sigma\mu} \langle A_{\sigma\mu} A_{\mu\rho} \rangle - \Phi^{\sigma\mu\mu\sigma} \langle A_{\mu\rho} A_{\sigma\mu} \rangle) + \sum_{\mu \neq \rho} (\Phi^{\mu\rho\rho\mu} \langle A_{\sigma\mu} A_{\mu\rho} \rangle - \Phi^{\rho\mu\mu\rho} \langle A_{\mu\rho} A_{\sigma\mu} \rangle), \tag{2}$$

where

$$\Phi^{\mu\nu\nu\mu} = \pi \sum_{\alpha\alpha'} \langle \alpha | h_{\mu\nu} | \alpha' \rangle \langle \alpha' | h_{\nu\mu} | \alpha \rangle P_{\alpha} \delta (\omega_{\mu\nu} + \omega_{\alpha\alpha'}). \tag{3}$$

This equation describes the relaxation of a system of molecules interacting with an electromagnetic field in free space. In particular, it can be used to study the spontaneous decay of a system of multilevel molecules (Section 2).

If we now regard the molecules and fields as the dynamic subsystem and the atoms and electrons on the cavity walls as the dissipative subsystem, then

$$E = \sum_{\nu} h_{\nu\nu} A_{\nu\nu} + \sum_{\mu<\nu} \omega_{\mu\nu} \left(a_{\mu\nu} a_{\nu\mu} + \frac{1}{2} \right) - \sum_{\mu<\nu} f_{\mu\nu} (a_{\mu\nu} A_{\mu\nu} + a_{\nu\mu} A_{\nu\mu}),$$

where $a_{\mu\nu}$ and $a_{\nu\mu} = a^{+}_{\mu\nu}$ are the photon creation and annihilation operators of the resonant mode, the coefficients $f_{\mu\nu}$ depend on the matrix elements of the dipole moments, and $\omega_{\mu\nu} = E_{\mu} - E_{\nu}$. The interaction Hamiltonian is

$$V = \sum_{\mu<\nu} (a_{\mu\nu} + a_{\nu\mu}) F_{\mu\nu},$$

where $F_{\mu\nu}$ describes the interaction of the field with the cavity walls.

The kinetic equations (1) then take the form

$$\langle \dot{a}_{mn} \rangle = i (\omega_{mn} a_{mn} - f_{mn} A_{mn}) - \gamma_{mn} \langle a_{mn} \rangle, \tag{4}$$

$$\langle \dot{A}_{\rho\sigma} \rangle = i \left[(h_{\sigma\sigma} - h_{\rho\rho}) A_{\rho\sigma} - \sum_{\nu\neq\sigma} f_{\sigma\nu} a_{\sigma\nu} A_{\rho\nu} + \sum_{\nu\neq\rho} f_{\rho\nu} a_{\nu\rho} A_{\nu\sigma} \right], \tag{5}$$

where

$$\gamma_{mn} = \pi \sum_{\alpha\alpha'} \langle \alpha | F_{mn} | \alpha' \rangle \langle \alpha' | F_{mn} | \alpha \rangle P_{\alpha} \delta (\omega_{mn} + \omega_{\alpha\alpha'}) \left[1 - \exp \left(- \frac{\omega_{mn}}{kT} \right) \right] \tag{5'}$$

and T is the temperature of the cavity walls. Equations (4) and (5) describe the behavior of a radiating system situated in a lossy cavity. Without dwelling on the general properties of these equations (discussed in [6]), we note that the system of equations (2)-(5) describes the general case of a real multilevel system. We now proceed to consider a system of three-level molecules.

If the energy difference between the first and second levels is ω_1, and that between the second and third is ω_2, then the Hamiltonian of the internal energy of the molecules is equal to

$$H = \sum_{j=1}^{N} H_j = \sum_{j=1}^{N} \left[\omega_1 I_{3j} + \frac{1}{3} \left(\omega_2 + \frac{\omega_1}{2} \right) (1 - 2K_{3j} - 2L_{3j}) \right], \tag{6}$$

where the operators I_{3j}, K_{3j}, and L_{3j} are those introduced in [1]. The Hamiltonian of the interaction with the field has the form

$$V = \sum_{j} (I_j^+ F_j^- + I_j^- F_j^+ + K_j^+ G_j^- + K_j^- G_j^+ + L_j^+ H_j^- + L_j^- H_j^+), \tag{7}$$

where

$$F_j^{\pm} = -(\mathbf{A}_{\omega_1}, \mathbf{e}^{\pm}_{(1)}), \quad G_j^{\pm} = -(\mathbf{A}_{\omega_2}, \mathbf{e}^{\pm}_{(2)}), \quad H_j^{\pm} = -(\mathbf{A}_{\omega_1+\omega_2}, \mathbf{e}^{\pm}_{(3)}),$$

$$\mathbf{e}^{\pm}_{(1)} = \mathbf{e}_1 \pm i\mathbf{e}_2, \quad \mathbf{e}^{\pm}_{(2)} = \mathbf{e}_3 \pm i\mathbf{e}_4, \quad \mathbf{e}^{\pm}_{(3)} = \mathbf{e}_5 \pm i\mathbf{e}_6,$$

$$\mathbf{e}_1 - i\mathbf{e}_2 = -\frac{i\omega_1}{c} \mathbf{d}_{12}, \quad \mathbf{e}_3 - i\mathbf{e}_4 = \frac{i\omega_2}{c} \mathbf{d}_{32}, \quad \mathbf{e}_5 - i\mathbf{e}_6 = \frac{i(\omega_1 + \omega_2)}{c} \mathbf{d}_{31},$$

$\mathbf{d}_{ik}$ are the matrix elements of the dipole moment of a molecule, $\mathbf{A}_{\omega_i}$ is the operator for the vector potential of a field of frequency ω_i at the system site, and the remaining operators [generators of SU(3)] are defined in [1]. In the case of a resonant cavity the notation

$$V = -\beta_1 (a_{12}^+ I^- + a_{12} I^+) - \beta_2 (a_{23}^+ K^+ + a_{23} K^-) - \beta_3 (a_{13}^+ L^+ + a_{13} L^-) \tag{8}$$

will also be used for V, where

$$\beta_1 = \sqrt{\frac{\omega_1}{2\tau}} (\mathbf{d}_{12}, \mathbf{e}_{\nu_1}), \quad \beta_2 = \sqrt{\frac{\omega_2}{2\tau}} (\mathbf{d}_{23}, \mathbf{e}_{\nu_2}), \quad \beta_3 = \sqrt{\frac{\omega_1 + \omega_2}{2\tau}} (\mathbf{d}_{13}, \mathbf{e}_{\nu_3}),$$

$\mathbf{e}_{\nu_i}$ is the polarization vector of a wave at frequency ν_i, ν_i are the eigenfrequencies of the cavity, and τ is the quantization volume of the field.

2. Spontaneous Emission into Free Space

According to Eqs. (2), (3), (6), and (7) the spontaneous radiation from a system of three-level molecules is described by the equations

$$\begin{aligned} \langle \dot{I}^+ \rangle &= i\omega_1 \langle I^+ \rangle + \gamma_1 \langle I^+ I_3 \rangle + {}^1/_2 (\gamma_2 + \gamma_3) \langle L^- K^+ \rangle, \\ \langle \dot{K}^+ \rangle &= -i\omega_2 \langle K^+ \rangle - \gamma_2 \langle K_3 K^+ \rangle - {}^1/_2 (\gamma_1 + \gamma_3) \langle L^+ I^+ \rangle, \\ \langle \dot{L}^+ \rangle &= -i(\omega_1 + \omega_2) \langle L^+ \rangle - \gamma_3 \langle L_3 L^+ \rangle + {}^1/_2 (\gamma_1 - \gamma_2) \langle I^- K^+ \rangle. \end{aligned} \tag{9}$$

$$\begin{aligned} \langle \dot{I}_3 \rangle &= -\gamma_1 \langle I^+ I^- \rangle + {}^1/_2 \gamma_2 \langle K^+ K^- \rangle - {}^1/_2 \gamma_3 \langle L^+ L^- \rangle, \\ \langle \dot{K}_3 \rangle &= -{}^1/_2 \gamma_1 \langle I^+ I^- \rangle + \gamma_2 \langle K^+ K^- \rangle + {}^1/_2 \gamma_3 \langle L^+ L^- \rangle, \\ \langle \dot{L}_3 \rangle &= -\langle \dot{I}_3 \rangle + \langle \dot{K}_3 \rangle, \end{aligned}$$

where γ_i is the natural linewidth of an isolated molecule for the i-th transition. Neglecting quantum fluctuations and proceeding to the real amplitudes, we obtain

$$\begin{aligned} \dot{r} &= \gamma_1 r r_3 + {}^1/_2 (\gamma_2 + \gamma_3) p l, \\ \dot{p} &= -{}^1/_2 (\gamma_1 + \gamma_3) r l - \gamma_2 p p_3, \\ \dot{l} &= {}^1/_2 (\gamma_1 - \gamma_2) p r - \gamma_3 l l_3, \\ \dot{r}_3 &= -\gamma_1 r^2 + {}^1/_2 \gamma_2 p^2 - {}^1/_2 \gamma_3 l^2, \\ \dot{p}_3 &= -{}^1/_2 \gamma_1 r^2 + \gamma_2 p^2 + {}^1/_2 \gamma_3 l^2, \\ \dot{l}_3 &= \dot{p}_3 - \dot{r}_3, \end{aligned} \tag{10}$$

where $r = \langle I^\pm \rangle$, $p = \langle K^\pm \rangle$, $l = \langle L^\pm \rangle$, $r_3 = \langle I_3 \rangle$, $p_3 = \langle K_3 \rangle$, and $l_3 = \langle L_3 \rangle$. There are three integrals of the motion for this system which facilitate the numerical integration:

$$\begin{aligned} l_3 &= p_3 - r_3, \\ f^2 &= r_3^2 + {}^1/_3 (p_3 + l_3)^2 + r^2 + p^2 + l^2, \\ g^3 &= -{}^2/_9 (p_3 + l_3)^3 + (2r_3^2 + 2r^2 - p^2 - l^2)(p_3 + l_3) + 3r_3 (p^2 - l^2) + 6rpl. \end{aligned} \tag{11}$$

We shall first examine the special case where $|r|$, $|p|$, $|l|$ have the maximum possible values at the initial time. Then Eq. (10) is solved analytically. It is easy to show that $0 \leqslant |r| \leqslant \sqrt{n_1 n_2}$, $0 \leqslant |p| \leqslant \sqrt{n_2 n_3}$, and $0 \leqslant |l| \leqslant \sqrt{n_1 n_3}$ always hold, where n_i are the populations of the corresponding levels. Using the integrals (11), the quantum mechanical average Casimir operators, it can be shown that if at the initial time $|r|$, $|p|$, and $|l|$ have the maximum possible values, then at any other time these equalities are valid:

$$|r(t)| = \sqrt{n_1(t)\, n_2(t)}, \quad |p(t)| = \sqrt{n_2(t)\, n_3(t)}, \quad |l(t)| = \sqrt{n_1(t)\, n_3(t)}.$$

Keeping in mind the equations

$$r_3=\frac{1}{2}(n_2-n_1),\quad p_3=\frac{1}{2}(n_2-n_3),\quad l_3=\frac{1}{2}(n_1-n_3);$$

the system (10) may be reduced to equations for the populations of an arbitrary pair of levels:

$$\dot{n}_1=-\gamma_1 n_1^2-(\gamma_1-\gamma_3)\,n_1n_3+\gamma_1 n_1 N,$$
$$\dot{n}_3=\gamma_2 n_3^2+(\gamma_2-\gamma_3)\,n_1n_3-\gamma_2 n_3 N.$$

For simplicity we shall examine two special cases:

1. If $\gamma_1=\gamma_2=\gamma_3=\gamma$, then

$$n_1(t)=\frac{n_{10}}{n_{10}+(N-n_{10})\,e^{-\gamma Nt}},\quad n_3(t)=\frac{n_{30}}{n_{30}+(N-n_{30})\,e^{\gamma Nt}},$$

where the subscript 0 denotes initial values.

2. If $\gamma_1=\gamma_2=\gamma$ and $\gamma_3=0$, then

$$n_1(t)=n_{10}\frac{a_3e^{-\gamma\alpha t}+b_3}{a_1+b_1e^{-\gamma\alpha t}},\quad n_3(t)=n_{30}\frac{a_1e^{\gamma\alpha t}+b_1}{a_3+b_3e^{\gamma\alpha t}},$$

where $a_{1,3}=(-N+2n_{10,30}+\alpha)$, $b_{1,3}=(N-2n_{10,30}+\alpha)$, and $\alpha=\sqrt{N^2-4n_{10}n_{30}}$.

The radiant intensities are

$$I_{2\to1}=\dot{n}_1=\frac{4n_{10}n_{20}\gamma\alpha^2e^{-\gamma\alpha t}}{(a_1+b_1e^{-\gamma\alpha t})^2},\quad I_{2\to1}(0)=\gamma n_{10}n_{20},$$
$$I_{3\to2}=-\dot{n}_3=\frac{4n_{20}n_{30}\gamma\alpha^2e^{\gamma\alpha t}}{(a_3+b_3e^{\gamma\alpha t})^2},\quad I_{3\to2}(0)=\gamma n_{20}n_{30}. \tag{12}$$

When

$$n_{10}\leqslant 0.5,\quad n_{30}\leqslant 0.5N \tag{13}$$

the intensities (12) have maxima at the points

$$t_{m2\to1}=\frac{1}{\alpha\gamma}\ln\frac{b_1}{a_1},\quad t_{m3\to2}=\frac{1}{\alpha\gamma}\ln\frac{a_3}{b_3},$$

given by

$$I_{m2\to1}=I_{m3\to2}={}^1\!/_4\,\alpha^2\gamma.$$

If conditions (13) are not satisfied, then the intensities decrease monotonically from their initial values.

Figures 1b and 2a illustrate Eqs. (12). We note that the case with maximum $|r|$, $|p|$, $|l|$ corresponds to a molecular state with the highest degree of coherence (cf. [3]). Such states decay the most rapidly. Figure 1 shows the intensity dependences for various values of γ_2/γ_1 when the levels are populated equally at the initial time with $\gamma_3=0$. It can be seen that an increase in this ratio leads to a rise in the intensity maxima and to a reduction in the duration of the responses. Figure 2 shows the dependence of the location of the intensity maxima on r_0, p_0, and l_0 for the case where $\gamma_1=\gamma_2=\gamma$, $\gamma_3=0$, and practically all the molecules are excited at the initial time. Clearly a reduction in p_0 causes the same shift in the $I_{2\to1}$ and $I_{3\to2}$ curves, and a reduction in r_0 and l_0 increases the distance between the maxima. The decay time of the system is inversely proportional to N in all cases, a consequence of the coherence of the radiators.

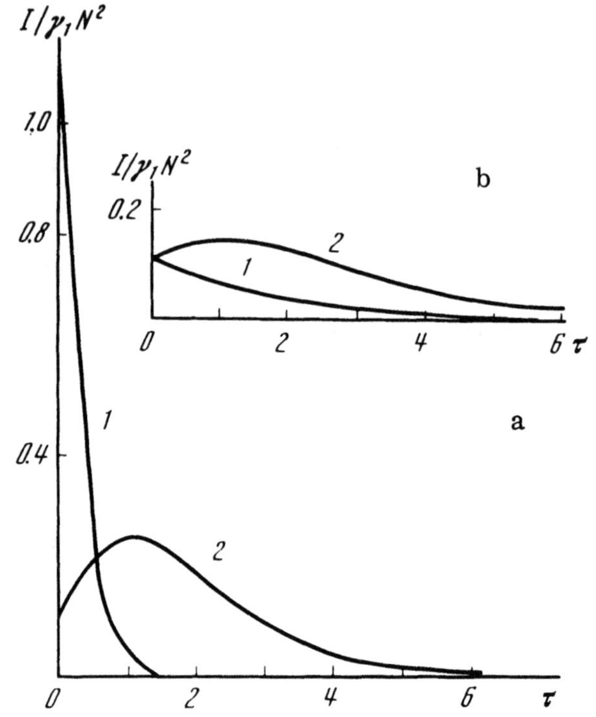

Fig. 1. The radiant intensities $\tau = t\gamma_1 N$; $n_{10} = n_{20} = n_{30} = N/3$; $r_0 = p_0 = l_0 = N/3$; $\gamma_3 = 0$. a) $\gamma_2/\gamma_1 = 10$; b) $\gamma_2/\gamma_1 = 1$. Curve 1) $I_{3\to 2}$; curve 2) $I_{2\to 1}$.

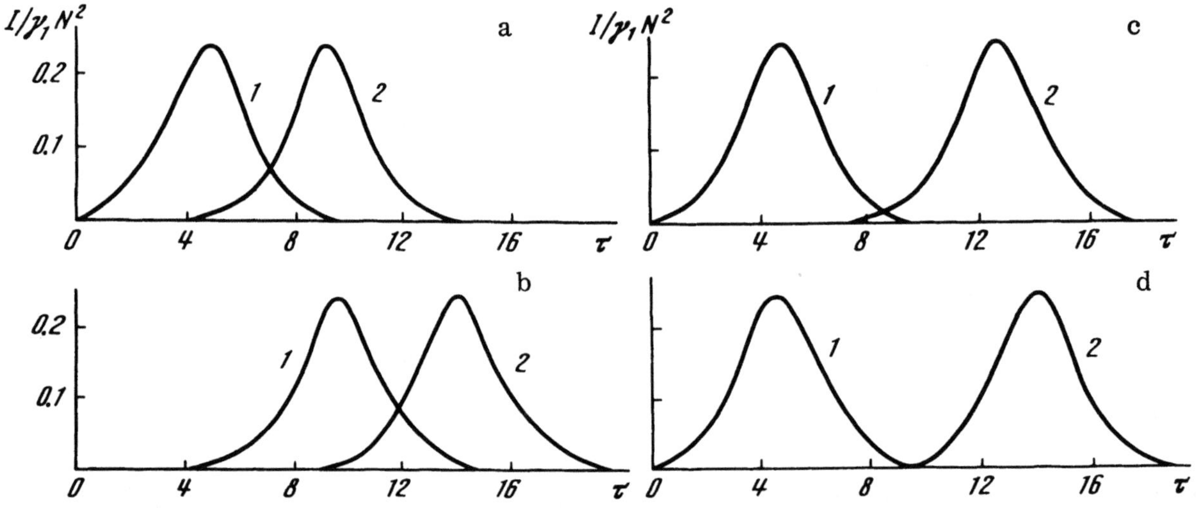

Fig. 2. Dependence of the location of the intensity maxima on r_0, p_0, l_0, and $\tau = t\gamma_1 N$. $n_{10} = n_{20} = 0.01N$; $n_{30} = 0.98N$. a) $r_0/N = 0.01$, $p_0/N = 0.1$, $l_0/N = 0.1$; b) $r_0/N = 0.01$, $p_0/N = 0.01$, $l_0/N = 0.1$; c) $r_0/N = 0.01$, $p_0/N = 0.1$, $l_0/N = 0.01$; d) $r_0/N = 0.001$, $p_0/N = 0.1$, $l_0/N = 0.01$. Curve 1) $I_{3\to 2}$; curve 2) $I_{2\to 1}$.

Thus, under coherent interaction conditions the effect of one radiative decay channel on another becomes very important, causing a change in the decay rate, shape, and time of appearance of the pulse.

3. Radiation in a Cavity

We shall consider a set of three-level molecules located in a resonator cavity whose eigenfrequencies ω_{ν_i} are close to those of the molecular transitions ω_i. According to Eqs.

(4)-(6) and (8), the dynamics of such a system is determined by the equations

$$\langle \dot{a}_{12}^{+}\rangle = i\omega_{\nu_1}\langle a_{12}^{+}\rangle - i\beta_1\langle I^{+}\rangle - \gamma_1\langle a_{12}^{+}\rangle,$$
$$\langle \dot{a}_{23}^{+}\rangle = i\omega_{\nu_2}\langle a_{23}^{+}\rangle - i\beta_2\langle K^{-}\rangle - \gamma_2\langle a_{23}^{+}\rangle,$$
$$\langle \dot{a}_{13}^{+}\rangle = i\omega_{\nu_3}\langle a_{13}^{+}\rangle - i\beta_3\langle L^{-}\rangle - \gamma_3\langle a_{13}^{+}\rangle,$$
$$\langle \dot{I}^{+}\rangle = i\omega_1\langle I^{+}\rangle + i2\beta_1\langle a_{12}^{+}I_3\rangle - i\beta_2\langle a_{23}L^{-}\rangle + i\beta_3\langle a_{13}^{+}K^{+}\rangle,$$
$$\langle \dot{K}^{+}\rangle = -i\omega_2\langle K^{+}\rangle - i\beta_1\langle a_{12}^{+}L^{+}\rangle + i2\beta_2\langle a_{23}K_3\rangle + i\beta_3\langle a_{13}I^{+}\rangle,$$
$$\langle \dot{L}^{+}\rangle = -i(\omega_1+\omega_2)\langle L^{+}\rangle - i\beta_1\langle a_{12}^{+}K^{+}\rangle + i\beta_2\langle d_{23}I^{-}\rangle + i2\beta_3\langle a_{13}L_3\rangle,$$

$$\langle \dot{I}_3\rangle = -i\beta_1\langle a_{12}^{+}I^{-} - a_{12}I^{+}\rangle + \frac{i}{2}\beta_2\langle a_{23}^{+}K^{+} - a_{23}K^{-}\rangle - \frac{i}{2}\beta_3\langle a_{13}^{+}L^{+} - a_{13}L^{-}\rangle,$$
$$\langle \dot{K}_3\rangle = -\frac{i}{2}\beta_1\langle a_{12}^{+}I^{-} - a_{12}I^{+}\rangle + i\beta_2\langle a_{23}^{+}K^{+} - a_{23}K^{-}\rangle + \frac{i}{2}\beta_3\langle a_{13}^{+}L^{+} - a_{13}L^{-}\rangle,$$
$$\langle \dot{L}_3\rangle = \langle \dot{K}_3\rangle - \langle \dot{I}_3\rangle,$$

where $\gamma_1 = \gamma_{12}$, $\gamma_2 = \gamma_{23}$, $\gamma_3 = \gamma_{13}$, and γ_{mn} is given by Eq. (5'). Introducing phenomenological values of the longitudinal and transverse relaxation times for each of the transitions, we find, after some simple transformations, that

$$\dot{z}_1 = \beta_1 r - \gamma_1 z_1, \quad \dot{z}_2 = \beta_2 p - \gamma_2 z_2, \quad \dot{z}_3 = \beta_3 l - \gamma_3 z_3,$$
$$\dot{r} = 2\beta_1 z_1 r + \beta_2 z_2 l + \beta_3 z_3 p - r/T_{2r},$$
$$\dot{p} = -\beta_1 z_1 l - 2\beta_2 z_2 p_3 - \beta_3 z_3 r - p/T_{2p},$$
$$\dot{l} = \beta_1 z_1 p - \beta_2 z_2 r - 2\beta_3 z_3 l_3 - l/T_{2l}, \tag{14}$$
$$\dot{r}_3 = -2\beta_1 z_1 r + \beta_2 z_2 p - \beta_3 z_3 l - (r_3 - r_3^0)/T_{1r},$$
$$\dot{p}_3 = -\beta_1 z_1 r + 2\beta_2 z_2 p + \beta_3 z_3 l - (p_3 - p_3^0)/T_{1p},$$
$$\dot{l}_3 = \beta_1 z_1 r + \beta_2 z_2 p + 2\beta_3 z_3 l - (l_3 - l_3^0)/T_{1l},$$

where $z_1 = \langle a_{12}^{+}\rangle$, $z_2 = \langle a_{23}^{+}\rangle$, $z_3 = \langle a_{13}^{+}\rangle$, and r_3^0, p_3^0, and l_3^0 are the equilibrium values of the corresponding quantities.

The quantities z_i are the amplitudes of the fields of the corresponding transitions. In the absence of relaxation, besides Eq. (11) there are some integrals of the motion which express conservation of the number of particles,

$$M_1 = r_3 + z_1^2 + (z_3^2 - z_2^2)/2, \qquad M_2 = p_3 - z_2^2 + (z_1^2 - z_3^2)/2.$$

Figures 3 and 4 show the results of a numerical integration of Eq. (14) when relaxation processes are absent ($N = 10^9$). In Fig. 4, $\beta_3 = 0$ and β_2/β_1 has various values. Clearly, the appearance of a second mode ω_2 leads to a strong change in the field amplitude of the first mode. The second mode itself has a fairly complicated time dependence which is explained by the strong coupling of both modes through the second level. The effect of the third mode on oscillations in the density of photons from the first two modes is illustrated in Fig. 4.

These numerical solutions illustrate the effect of the initial conditions on the structure of the field oscillations in a three-mode resonator. These data are already an indication of the complex dynamics of real systems of this type. We note that one or another set of initial conditions may be established with the aid of input radiation. Thus, such systems make it possible in principle to modulate optical fields and control them. We note that a detailed analysis of coherent control systems will require the solution of systems of equations such as (14) taking into account the effect of pumping under equilibrium initial conditions.

4. Propagation and Interaction of Ultrashort Pulses in a Three-Level Resonant Medium

We now consider the propagation of ultrashort pulses in a resonant medium under coherent interaction conditions when the width of the pulse spectra is much greater than the homogeneous broadening width (the two-level case was first examined in [7]).

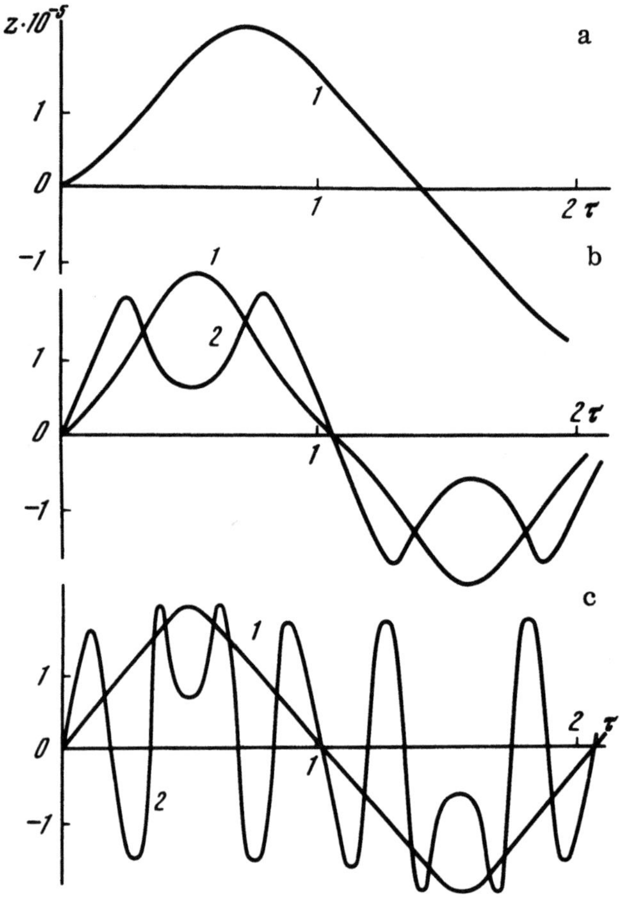

Fig. 3. Effect of the second mode on the amplitude of the first mode. $\tau = t\beta_1 N \cdot 10^{-5}$; $n_{10} = n_{20} = n_{30} = N/3$; $r_0 = p_0 = l_0 = N/3$. a) $\beta_2 = 0$, $\beta_3 = 0$; b) $\beta_2/\beta_1 = 3$, $\gamma_3 = 0$; c) $\beta_2/\beta_1 = 10$, $\beta_3 = 0$. Curve 1) z_1; curve 2) z_2.

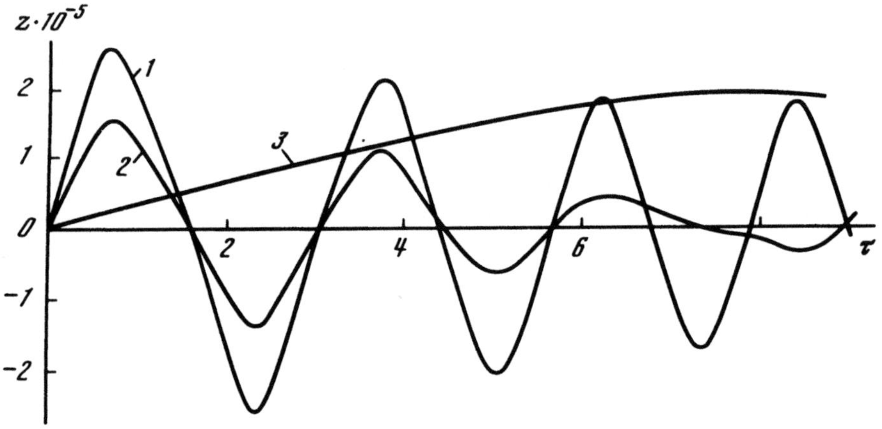

Fig. 4. Effect of the third mode. $\tau = t\beta_1 N \cdot 10^{-5}$; $n_{10} = n_{20} = n_{30} = N/3$; $r_0 = p_0 = l_0 = N/3$; $\beta_2/\beta_1 = 1$; $\beta_3/\beta_1 = 0.1$. Curves 1, 2, 3) $z_{1,2,3}$, respectively.

We shall first obtain a general system of equations describing the propagation of resonant pulses with frequencies ω_{12}^0, ω_{23}^0, and ω_{13}^0 ($\omega_{13}^0 = \omega_{12}^0 + \omega_{23}^0$) equal to the central frequencies of the transitions. In our discussion we shall use a semiclassical approach in which the medium is described quantum mechanically, and the field, classically. Then the effects associated with spontaneous emission are not taken into account and may be neglected in this problem.

The unperturbed Hamiltonian of the system can be written as in Eq. (6). In the dipole approximation the Hamiltonian of the interaction of the molecules and field has the form

$$H_1 = -p_1(E_-^{12}I_+ + E_+^{12}I_-) - p_2(E_-^{23}K_- + E_+^{23}K_+) - p_3(E_-^{13}L_- + E_+^{13}L_+), \tag{15}$$

where the circularly polarized fields are given by

$$\mathbf{E}^{ik}(z,\ t) = \varepsilon^{ik}(z,\ t)\{\mathbf{i}\cos[\omega_{ik}t - k_{ik}z - \Phi_{ik}(z,\ t)] + \mathbf{j}\sin[\omega_{ik}t - k_{ik}z - \Phi_{ik}(z,\ t)]\},$$
$$i,\ k = 1,\ 2,\ 3, \quad i<k, \quad E_\pm^{ik} = E_x^{ik} \pm iE^{ik},$$

and $p_i/\sqrt{2}$ are the dipole moments of the transitions. In the following we shall solve this problem without taking phase modulation into account (Φ_{ik} = const).

The equations of motion for the average values of the polarization vectors have the form

$$\begin{aligned}
&\frac{d\tilde I_x}{dt} = -\Delta\omega_1\tilde I_y - \frac{p_2}{\hbar}\tilde L_y\varepsilon^{23} - \frac{p_3}{\hbar}\tilde K_y\varepsilon^{13},\\
&\frac{d\tilde I_y}{dt} = \Delta\omega_1\tilde I_x + \frac{2p_1}{\hbar}\varepsilon^{12}I_3 - \frac{p_2}{\hbar}\tilde L_x\varepsilon^{23} + \frac{p_3}{\hbar}\tilde K_x\varepsilon^{13},\\
&\frac{dI_3}{dt} = -\frac{2p_1}{\hbar}\varepsilon^{12}\tilde I_y - \frac{p_2}{\hbar}\varepsilon^{23}\tilde K_y + \frac{p_3}{\hbar}\varepsilon^{13}\tilde I_y,\\
&\frac{d\tilde K_x}{dt} = \Delta\omega_2\tilde K_y + \frac{p_1}{\hbar}\varepsilon^{12}\tilde L_y - \frac{p_3}{\hbar}\varepsilon^{13}\tilde I_y,\\
&\frac{d\tilde K_y}{dt} = -\Delta\omega_2\tilde K_x - \frac{p_1}{\hbar}\varepsilon^{12}\tilde L_x + \frac{2p_2}{\hbar}\varepsilon^{23}K_3 + \frac{p_3}{\hbar}\varepsilon^{13}\tilde I_x,\\
&\frac{dK_3}{dt} = -\frac{p_1}{\hbar}\varepsilon^{12}\tilde I_y - \frac{2p_2}{\hbar}\varepsilon^{23}\tilde K_y - \frac{p_3}{\hbar}\varepsilon^{13}\tilde L_y,\\
&\frac{d\tilde L_x}{dt} = (\Delta\omega_1 + \Delta\omega_2)\tilde L_y + \frac{p_1}{\hbar}\varepsilon^{12}\tilde K_y + \frac{p_2}{\hbar}\varepsilon^{23}\tilde I_y,\\
&\frac{d\tilde L_y}{dt} = -(\Delta\omega_1 + \Delta\omega_2)\tilde L_x - \frac{p_1}{\hbar}\varepsilon^{12}\tilde K_x + \frac{p_2}{\hbar}\varepsilon^{23}\tilde I_x + \frac{2p_3}{\hbar}\varepsilon^{13}L_3,\\
&L_3 = K_3 - I_3.
\end{aligned} \tag{16}$$

Here the notation

$$\begin{aligned}
&\langle I_\pm\rangle = \tilde I_\pm \exp[\pm i(\omega_{12}t - k_{12}z)],\\
&\langle L_\pm\rangle = \tilde L_\pm \exp[\mp i(\omega_{13}t - k_{13}z)],\\
&\langle K_\pm\rangle = \tilde K_\pm \exp[\mp i(\omega_{23}t - k_{23}z)],\\
&\Delta\omega_1 = \omega_{12}^0 - \omega_{12}, \qquad \Delta\omega_2 = \omega_{23}^0 - \omega_{23}
\end{aligned}$$

is used and the transition is made to the x- and y-components of the polarization vectors ($I_\pm = I_x \pm iI_y$, etc.). In deriving Eqs. (16) the condition

$$\mathbf{k}_{12} + \mathbf{k}_{23} = \mathbf{k}_{13} \tag{17}$$

was applied.

From Maxwell's equations for the circularly polarized waves,

$$\frac{\partial^2 E_+^{ik}}{\partial z^2} = \frac{\eta_{ik}^2}{c^2}\frac{\partial^2 E_+^{ik}}{\partial t^2} + \frac{4\pi}{c^2}\frac{\partial^2 P_+^{ik}}{\partial t^2},$$

where P_+^{ik} are the polarizations averaged over the inhomogeneously broadened line,

$$P_+^{12}=\int_{-\infty}^{+\infty} g_{12}(\Delta\omega_1)\,[2Np_1(\langle I_x\rangle+i\langle I_y\rangle)]\,d(\Delta\omega_1),$$
$$\int g_{ik}(\Delta\omega)\,d(\Delta\omega)=1,\ \text{etc.,} \tag{18}$$

and η_{ik} is the nonresonant part of the refractive index, we obtain the following in the envelope approximation for $\omega_{12}\neq\omega_{23}\neq\omega_{13}$:

$$\frac{\partial\varepsilon_{12}}{\partial z}=-\frac{\eta_{12}}{c}\frac{\partial\varepsilon_{12}}{\partial t}+\frac{2\pi\omega_{12}}{\eta_{12}c}\int_{-\infty}^{+\infty} g_{12}(\Delta\omega_1)\,2Np_1\tilde{I}_y\,d(\Delta\omega_1),$$
$$\frac{\partial\varepsilon_{23}}{\partial z}=-\frac{\eta_{23}}{c}\frac{\partial\varepsilon_{23}}{\partial t}-\frac{2\pi\omega_{23}}{\eta_{23}c}\int_{-\infty}^{+\infty} g_{23}(\Delta\omega_2)\,2Np_2\tilde{K}_y\,d(\Delta\omega_2), \tag{19}$$
$$\frac{\partial\varepsilon_{13}}{\partial z}=-\frac{\eta_{13}}{c}\frac{\partial\varepsilon_{13}}{\partial t}-\frac{2\pi\omega_{12}}{\eta_{13}c}\int_{-\infty}^{+\infty} g_{13}(\Delta\omega_3)\,2Np_3\tilde{L}_y\,d(\Delta\omega_3).$$

When the frequencies are equal, we find for $\omega_{12}=\omega_{23}$ (equidistant case) in place of Eq. (19) that

$$\frac{\partial\varepsilon}{\partial z}=-\frac{\eta_{12}}{c}\frac{\partial\varepsilon}{\partial t}+\frac{4\pi\omega_{12}}{\eta_{12}c}N\left[\int_{-\infty}^{+\infty} g_{12}(\Delta\omega_1)\,p_1\tilde{I}_y\,d(\Delta\omega_1)-\int_{-\infty}^{+\infty} g_{23}(\Delta\omega_2)\,p_2\tilde{K}_y\,d(\Delta\omega_2)\right]; \tag{20}$$

for $\omega_{12}=\omega_{13}$ (double upper level) that

$$\frac{\partial\varepsilon}{\partial z}=-\frac{\eta_{12}}{c}\frac{\partial\varepsilon}{\partial t}+\frac{4\pi\omega_{12}}{\eta_{12}c}N\left[\int_{-\infty}^{+\infty} g_{12}(\Delta\omega_1)\,p_1\tilde{I}_y\,d(\Delta\omega_1)-\int_{-\infty}^{+\infty} g_{13}(\Delta\omega_3)\,p_3\tilde{L}_y\,d(\Delta\omega_3)\right]; \tag{21}$$

and for $\omega_{23}=\omega_{13}$ (double lower level) that

$$\frac{\partial\varepsilon}{\partial z}=-\frac{\eta_{13}}{c}\frac{\partial\varepsilon}{\partial t}-\frac{4\pi\omega_{13}}{\eta_{12}c}N\left[\int_{-\infty}^{+\infty} g_{23}(\Delta\omega_2)\,p_2\tilde{K}_y\,d(\Delta\omega_2)+\int_{-\infty}^{+\infty} g_{13}(\Delta\omega_3)\,p_3\tilde{L}_y\,d(\Delta\omega_3)\right]. \tag{22}$$

The system of equations (16) with (19)-(22) describes the propagation of pulses under coherent interaction conditions. We now proceed to solve it.

The Equidistant Case ($\omega_{12}=\omega_{23}$). We assume that the 1−3 transition is forbidden ($p_3 = 0$). In this case condition (17) is automatically satisfied. Then the system (16) takes the form

$$\begin{aligned}
\frac{d\tilde{I}_x}{dt}&=-\Delta\omega_1\tilde{I}_y-\frac{p_2}{\hbar}\tilde{L}_y\varepsilon,\\
\frac{d\tilde{I}_y}{dy}&=\Delta\omega_1\tilde{I}_x+\frac{2p_1}{\hbar}\varepsilon I_3-\frac{p_2}{\hbar}L_x\varepsilon,\\
\frac{dI_3}{dt}&=-\frac{2p_1}{\hbar}\varepsilon I_y-\frac{p_2}{\hbar}\varepsilon\tilde{K}_y,\\
\frac{d\tilde{K}_x}{dt}&=\Delta\omega_2\tilde{K}_y+\frac{p_1}{\hbar}\varepsilon\tilde{L}_y,\\
\frac{d\tilde{K}_y}{dt}&=-\Delta\omega_2\tilde{K}_x-\frac{p_1}{\hbar}\varepsilon\tilde{L}_x+\frac{2p_2}{\hbar}\varepsilon K_3,\\
\frac{dK_3}{dt}&=-\frac{p_1}{\hbar}\varepsilon\tilde{I}_y-\frac{2p_2}{\hbar}\varepsilon\tilde{K}_y,\\
\frac{d\tilde{L}_x}{dt}&=(\Delta\omega_1+\Delta\omega_2)\tilde{L}_y+\frac{p_1}{\hbar}\varepsilon\tilde{K}_y+\frac{p_2}{\hbar}\varepsilon\tilde{I}_y,\\
\frac{d\tilde{L}_y}{dt}&=-(\Delta\omega_1+\Delta\omega_2)\tilde{L}_x-\frac{p_1}{\hbar}\varepsilon\tilde{K}_x+\frac{p_2}{\hbar}\varepsilon\tilde{I}_x.
\end{aligned} \tag{23}$$

The solution of this system for an exact resonance ($\Delta\omega_1 = \Delta\omega_2 = 0$) with initial conditions

$$I_x(0) = I_y(0) = K_x(0) = K_y(0) = L_x(0) = L_y(0) = 0,$$
$$I_3(0) = A = \frac{1}{2}\frac{(N_2 - N_1)}{N_1 + N_2 + N_3}, \qquad K_3(0) = B = \frac{1}{2}\frac{(N_2 - N_3)}{N_1 + N_2 + N_3}$$

(N_i is the population of the i-th level at the initial time) has the form

$$\begin{aligned}
&\tilde{I}_x = 0,\\
&\tilde{I}_y = -C_1 \sin\frac{p\varphi}{2} - C_2 \sin p\varphi,\\
&I_3 = -\frac{C_1}{p}\cos\frac{p\varphi}{2} - \frac{2+\gamma^2}{2p} C_2 \cos p\varphi + \frac{A\gamma^2(2\gamma^2-1) + B\gamma^2(2-\gamma^2)}{2p^4},\\
&\tilde{K}_y = \frac{C_1}{\gamma}\sin\frac{p\varphi}{2} - \gamma C_2 \sin p\varphi,\\
&K_3 = \frac{C_1}{p}\cos\frac{p\varphi}{2} - \frac{2\gamma^2+1}{2p} C_2 \cos p\varphi + \frac{A(2\gamma^2-1) + B(2-\gamma^2)}{2p^4},\\
&\tilde{L}_x = -\frac{1-\gamma^2}{\gamma p} C_1 \cos\frac{p\varphi}{2} + \frac{\gamma}{p} C_2 \cos p\varphi + \frac{A\gamma(2\gamma^2-1) + B\gamma(2-\gamma^2)}{p^4},\\
&\tilde{K}_x = 0,\\
&\tilde{L}_y = 0;
\end{aligned} \tag{24}$$

where $C_1 = 2\frac{(B-A)\gamma^2}{p^3}$; $C_2 = -\frac{(A+\gamma^2 B)}{p^3}$, $\varphi = \frac{2p_1}{\hbar}\int_{-\infty}^{t}\varepsilon dt$; $\gamma = \frac{p_2}{p_1}$, $p = \sqrt{1+\gamma^2}$. If all the molecules are at the lower level at the initial time, then $A = -1/2$, $B = 0$ and $C_1 = \gamma^2/p^3$, $C_2 = 1/2p^3$.

Now, integrating Eq. (20), we find that

$$\frac{\partial\theta}{\partial z} = \frac{8\pi N\omega p_1^2}{\eta c\hbar}\int_{-\infty}^{+\infty}\left[\int_{-\infty}^{+\infty} g_{12}(\Delta\omega_1)\tilde{I}_y d(\Delta\omega_1) - \int_{-\infty}^{+\infty}\gamma g_{23}(\Delta\omega_2)\tilde{K}_y d(\Delta\omega_2)\right]dt, \tag{25}$$

where $\theta = \varphi(\infty) = \frac{2p_1}{\hbar}\int_{-\infty}^{+\infty}\varepsilon dt$. From the system (16) we have in our case that

$$\begin{aligned}
\tilde{I}_y &= -\frac{d\tilde{I}_x}{dt}\frac{1}{\Delta\omega_1} - \frac{p_2}{\hbar\Delta\omega_1}\varepsilon\tilde{L}_y,\\
\tilde{K}_y &= \frac{d\tilde{K}_x}{dt}\frac{1}{\Delta\omega_2} - \frac{p_1\varepsilon}{\hbar\Delta\omega_2}\tilde{L}_y,
\end{aligned} \tag{26}$$

where at time T after passage of the pulse

$$\begin{aligned}
\tilde{I}_x(\Delta\omega, z, T+t) &= \tilde{I}_x(\Delta\omega, z, T)\cos\Delta\omega t - \tilde{I}_y(\Delta\omega, z, T)\sin\Delta\omega t,\\
\tilde{K}_x(\Delta\omega, z, T+t) &= \tilde{K}_x(\Delta\omega, z, T)\cos\Delta\omega t + \tilde{K}_y(\Delta\omega, z, T)\sin\Delta\omega t.
\end{aligned} \tag{27}$$

Substituting Eqs. (24), (26), and (27) in Eq. (25), and using the relations

$$\lim_{t\to\infty}\frac{\sin xt}{x} = \pi\delta(x),$$
$$\lim_{t\to\infty}\frac{\cos xt}{x} = \frac{1}{x} - P\left(\frac{1}{x}\right),$$

we find the area theorem:

$$\frac{d\theta}{dz} = \frac{8\pi^2 N\omega p_1^2}{\eta c\hbar} g_{12}(0)\left[-C_1\sin\frac{p\theta}{2} - C_2\sin p\theta\right] + \frac{8\pi^2 N\omega p_1^2}{\eta c\hbar} g_{23}(0)\left[-C_1\sin\frac{p\theta}{2} + \gamma^2 C_2 \sin p\theta\right]. \tag{28}$$

In two special cases (all the molecules initially in the lower level and $\gamma = 0$; and all molecules in the second level and $\gamma \to \infty$), Eq. (28) transforms to McCall and Hahn's result [7] for the two-level problem.

From Eq. (28) it is easy to obtain Beer's absorption law for weak fields,

$$\frac{\partial W}{\partial t} = W_0 e^{-\alpha z},$$

where

$$W = \frac{\eta c}{8\pi} \int_{-\infty}^{+\infty} \varepsilon^2 (z, t)\, dt;$$

$$\alpha = \frac{8\pi^2 N \omega p_1^2}{\eta c \hbar} p \left[g_{12}(0)(C_1 + 2C_2) + g_{23}(0)(C_1 - 2\gamma^2 C_2)\right].$$

Integrating Eq. (28) and assuming $g_{12}(0) = g_{23}(0)$ (which can easily be shown for the case of Doppler broadening), we obtain

$$kz = \frac{1}{4C_1^2 - 4C_2^2(1-\gamma^2)^2}\left[-2C_1 \ln|\tan(p\theta/4)| - 2C_2(1-\gamma^2)\ln\frac{|-2C_1 - 2C_2(1-\gamma^2)\cos(p\theta/2)|}{|\sin(p\theta/2)|}\right]. \quad (29)$$

An examination of Eq. (29) shows that, depending on the initial populations and ratio of the transition strengths, the following (four) cases are possible:

(a) $(\gamma^2 - 1)C_2 > C_1$ or $A > B\gamma^2$,
$(1-\gamma^2)C_2 > C_1$ $\quad A(3\gamma^2 - 1) > B\gamma^2(3-\gamma^2)$.

The area theorem for this case is illustrated in Fig. 5a. Here the stationary pulse is a 2π - pulse.

(b) $(\gamma^2 - 1)C_2 > C_1$ or $A > B\gamma^2$,
$(1-\gamma^2)C_2 < C_1$ $\quad A(3\gamma^2 - 1) < B\gamma^2(3-\gamma^2)$

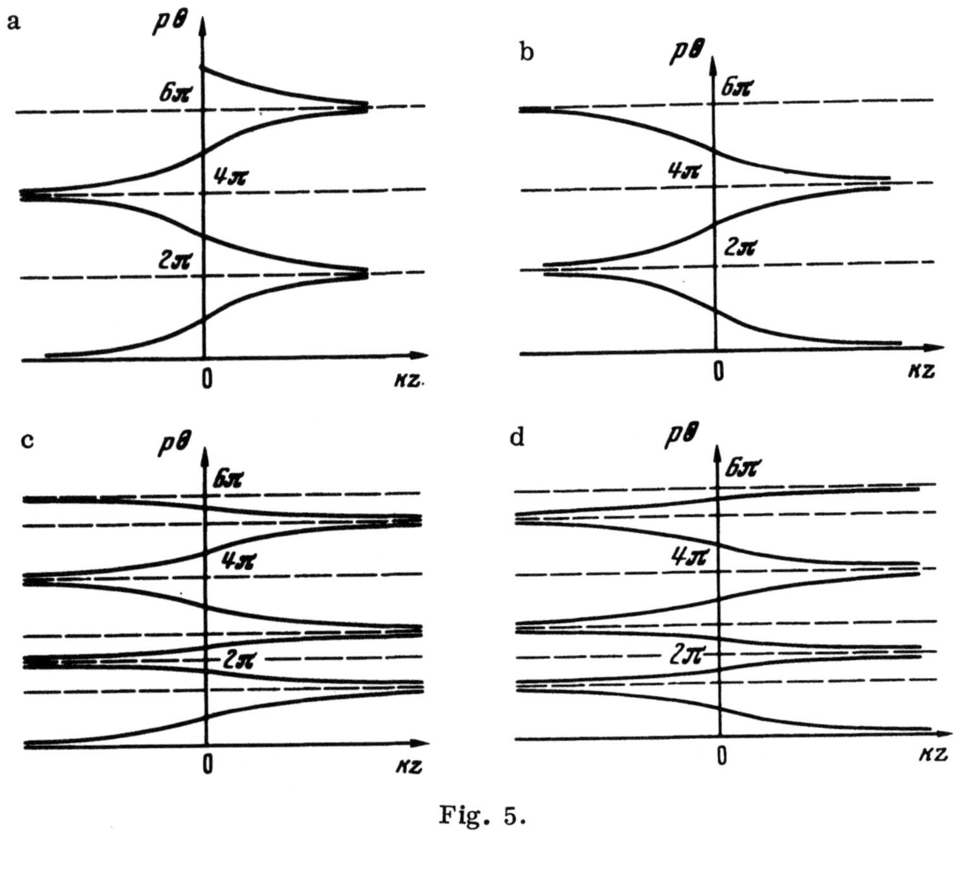

Fig. 5.

(cf. Fig. 5b). In this case the stationary points are given by the condition

$$C_1 + C_2(1-\gamma^2)\cos(p\theta/2) = 0. \tag{30}$$

Cases (a) and (b) refer to an amplifying medium.

$$\text{(c)}\quad \begin{array}{lll} (\gamma^2-1)\,C_2 < C_1 & \text{or} & A < B\gamma^2, \\ (1-\gamma^2)\,C_2 < C_1 & & A(3\gamma^2-1) < B\gamma^2(3-\gamma^2). \end{array}$$

Here the stationary pulse is a 4π-pulse (cf. Fig. 5c).

$$\text{(d)}\quad \begin{array}{lll} (\gamma^2-1)\,C_2 < C_1 & \text{or} & A < B\gamma^2, \\ (1-\gamma^2)\,C_2 > C_1 & & A(3\gamma^2-1) > B\gamma^2(3-\gamma^2) \end{array}$$

(cf. Fig. 5d). Here the unstable equilibrium points are also found using Eq. (30). Cases (c) and (d) refer to an absorbing medium. Case (d) corresponds to the two-level problem when $\gamma = 0$; then the asymptotes pass through the point $\theta = 2\pi(2n-1)$.

We now introduce the equation for the envelope of the stationary pulse which follows from Eq. (20):

$$\begin{aligned} \frac{\partial\varepsilon}{\partial\tau} = \frac{\alpha p}{4\hbar}\Big\{ & C_1\left(p_1+\frac{p_2}{\gamma}\right)\frac{2}{r}\left[\lambda r-\lambda^2-e^2-m\varepsilon^2+(r-2\lambda)\sqrt{e+m\varepsilon^2}\right]^{1/2}+ \\ & +C_2(p_1-\gamma p_2)\frac{4}{r}\Big[\lambda r-\lambda^2-e^2-\frac{4}{r^2}(\lambda r-\lambda^2-e^2)^2-\frac{4}{r^2}e(r-2\lambda)^2- \\ & -m\varepsilon^2-\frac{4}{r^2}m\varepsilon^2(r-2\lambda)^2-2(\lambda r-\lambda^2-e^2)m\varepsilon^2+ \\ & +\sqrt{e+m\varepsilon^2}\left(r-2\lambda+2(\lambda r-\lambda^2-e^2)(r-2\lambda)-2(r+2\lambda)m\varepsilon^2\right)\Big]^{1/2}\Big\}, \end{aligned}$$

where

$$\lambda = (\gamma^2-1)\,C_2-2C_1, \qquad e=\lambda^2, \qquad m=4C_2(\gamma^2-1)/\alpha,$$

$$\alpha = \frac{16\pi\hbar\omega N}{\eta c p\,(1/v-\eta/c)}, \qquad r=2C_2(\gamma^2-1), \qquad \tau = t - z/v.$$

For a system in a pure quantum state at t = 0 the study of the stationary pulses, or solitons, is much simpler. Then the elements of the density matrix for the medium are related by

$$|\rho_{ij}| = \sqrt{\langle\rho_{ii}\rangle\langle\rho_{jj}\rangle}. \tag{31}$$

Setting

$$\tilde{I}_y = -x_1x_2, \qquad \tilde{K}_y = x_2x_3, \qquad \tilde{L}_x = -x_1x_3$$

according to Eq. (31), it is easy to find from Eqs. (20) and (23) (see also [2]) that

$$\begin{aligned} \dot{x}_1 &= -2yx_2, \\ \dot{x}_2 &= 2y(x_1-\gamma x_3), \\ \dot{x}_3 &= 2\gamma y x_2, \\ \dot{y} &= x_1x_2+\gamma x_2x_3, \end{aligned}$$

where

$$y=\frac{p_1}{2\hbar}\varepsilon\Omega^{-1}, \quad \Omega^2=\frac{2\pi N p_1^2\omega_{12}}{\eta_{12}^2\hbar\,(c/v\eta_{12}-1)},$$

v is the group velocity of the stationary pulse, and the time is measured in units of Ω^{-1}.

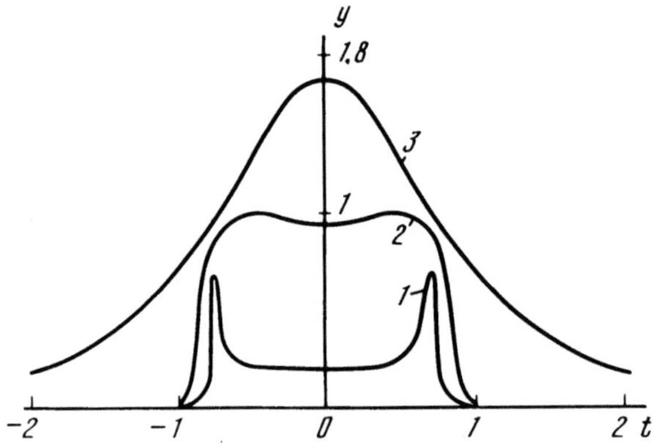

Fig. 6. Shape of the stationary pulse in an equidistant system with $x_2(0) = x_3(0) = 0$. Curve 1) $\gamma = 0.1$; 2) $\gamma = 0.5$; 3) $\gamma = 1$.

Introducing $\tau = \int_{-\infty}^{t} y dt$ and $p = (1 + \gamma^2)^{1/2}$, we have

$$
\begin{aligned}
x_1 &= C_1 \cos 2p\tau + C_2 \sin 2p\tau + C_3, \\
x_2 &= -pC_2 \cos 2p\tau + pC_1 \sin 2p\tau, \\
x_3 &= -\gamma C_1 \cos 2p\tau - \gamma C_2 \sin 2p\tau + C_3/\gamma, \\
y &= \pm\{{}^1/_2(\gamma^2 - 1) C_1 C_2 \sin 4p\tau + {}^1/_2 [C_2^2(\gamma^2 - 1) + C_1^2(1-\gamma^2)] \sin^2 2p\tau - \\
&\quad - 2C_2C_3 \sin 2p\tau - 2C_1C_3 \cos 2p\tau + C_4\}^{1/2}.
\end{aligned} \tag{32}
$$

We now consider Eq. (32) for various initial conditions. If at the initial time all the molecules are in the lower level and $y(0) = 0$, then (a study of this case was presented in [2])

$$
\begin{gathered}
x_1 = \frac{1}{p^2}(\cos 2p\tau + \gamma^2), \qquad x_2 = \frac{1}{p} \sin 2p\tau, \qquad x_3 = -\frac{\gamma}{p^2}(\cos 2p\tau - 1), \\
y = \frac{\sqrt{2} \sin p\tau}{p^2}[2\gamma^2 + (1-\gamma^2)\cos^2 p\tau]^{1/2} = \frac{d\tau}{dt}, \qquad t = -\frac{\sqrt{2}}{4} \ln \frac{[2\gamma^2 + (1-\gamma^2)\cos^2 p\tau]^{1/2} + p\cos p\tau}{[2\gamma^2 + (1-\gamma^2)\cos^2 p\tau]^{1/2} - p\cos p\tau}.
\end{gathered} \tag{33}
$$

From Eq. (33) it follows that for $\gamma < 1/\sqrt{3}$ the pulse is double-humped and for $\gamma > 1/\sqrt{3}$, it is single-humped (see Fig. 6).

If all the molecules are in the middle level when $t = 0$, then

$$
y = \frac{2}{p}\sqrt{\frac{\gamma^2 - 1}{2}} \frac{\exp[\sqrt{2}(p-1)t]}{1 + \exp[2\sqrt{2}(p-1)t]}.
$$

The pulse shape as a function of γ for this case is shown in Fig. 7.

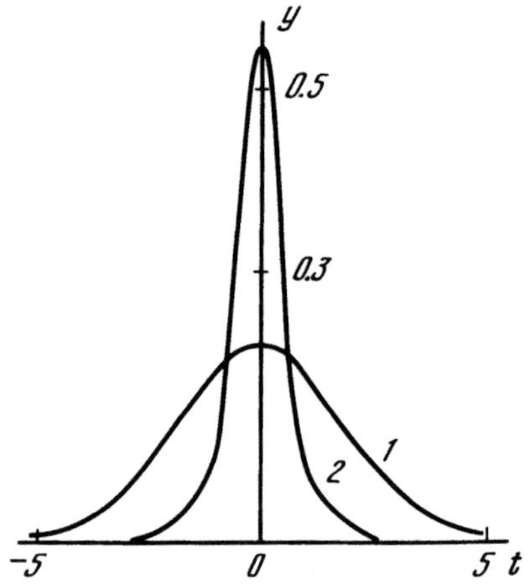

Fig. 7. The shape of the stationary pulse in an equidistant system when the middle level is populated. Curve 1) $\gamma = 1.1$; curve 2) $\gamma = 2$.

The Nonequidistant Case ($\omega_{12} \neq \omega_{23}$). As before, we shall assume the 1−3 transition to be forbidden. We shall find the form of the stationary as well as the propagating (through the medium) pulses with frequencies ω_{12} and ω_{23} for a system in a pure state at the initial time.

As before, we find

$$\begin{aligned} \dot{x}_1 &= -2y_1x_2, \\ \dot{x}_2 &= 2y_1x_1 - 2y_2x_3, \\ \dot{x}_3 &= 2y_2x_2, \\ \dot{y}_1 &= x_1x_2, \\ \dot{y}_2 &= \lambda^2x_2x_3, \end{aligned} \tag{34}$$

where

$$y_{1,2} = \frac{p_{1,2}}{2\hbar}\varepsilon_{12,23}\Omega^{-1}, \qquad \Omega^2 = \frac{2\pi N p_1^2\omega_{12}}{\eta_{12}^2\hbar\,(c/v\eta_{12}-1)}, \qquad \lambda = \frac{\eta_{12}p_2}{\eta_{23}p_1}\sqrt{\frac{\omega_{23}\,(c-v\eta_{12})}{\omega_{12}\,(c-v\eta_{23})}}.$$

Solving Eq. (34), we find

$$\begin{aligned} y_1 &= C_1\sin\sqrt{2}\tau + C_2\cos\sqrt{2}\tau, \\ x_1 &= \sqrt{2}C_1\cos\sqrt{2}\tau - \sqrt{2}C_2\sin\sqrt{2}\tau, \\ y_2 &= C_3\sinh\sqrt{2}\lambda\tau + C_4\cosh\sqrt{2}\lambda\tau, \\ x_3 &= \frac{\sqrt{2}C_3}{\lambda}\cosh\sqrt{2}\lambda\tau + \frac{\sqrt{2}C_4}{\lambda}\sinh\sqrt{2}\lambda\tau, \end{aligned}$$

where

$$\tau = \int_{-\infty}^{t} x_2\,dt \quad \text{for} \quad x_2 > 0,$$

$$\tau = \int_{\infty}^{t} x_2\,dt \quad \text{for} \quad x_2 < 0,$$

and the constants C_i are determined from the initial conditions. For x_2 we have

$$x_2 = \frac{d\tau}{dt} = \pm\Big[2C_1C_2\sin 2\sqrt{2}\tau + 2\,(C_1^2 - C_2^2)\sin^2\sqrt{2}\tau - \\ -\frac{2}{\lambda^2}C_3C_4\sinh 2\sqrt{2}\lambda\tau - \frac{2}{\lambda^2}(C_4^2 + C_3^2)\sinh^2\sqrt{2}\lambda\tau + C_5\Big]^{1/2}.$$

The case in which all the molecules are in the lower level was examined in [2]. In that case

$$\begin{aligned} y_1 &= \frac{1}{\sqrt{2}}\sin\sqrt{2}\,\tau, \\ x_1 &= \cos\sqrt{2}\,\tau, \\ y_2 &= y_0\cosh\sqrt{2}\,\lambda\tau, \\ x_3 &= \frac{\sqrt{2}\,y_0}{\lambda}\sinh\sqrt{2}\,\lambda\tau, \\ x_2 &= \frac{d\tau}{dt} = \pm\left(\sin^2\sqrt{2}\tau - \frac{2y_0^2}{\lambda^2}\sinh^2\sqrt{2}\,\lambda\tau\right)^{1/2}, \end{aligned} \tag{35}$$

where y_0 is a given initial field at the $2 \leftrightarrow 3$ transition. From Eq. (35) it follows that stationary propagation of the pulse is possible only when $y_0 < 1/\sqrt{2}$.

It is easily shown that if

$$\sinh^2\frac{\pi}{2}\lambda < \frac{\lambda^2}{2y_0^2},$$

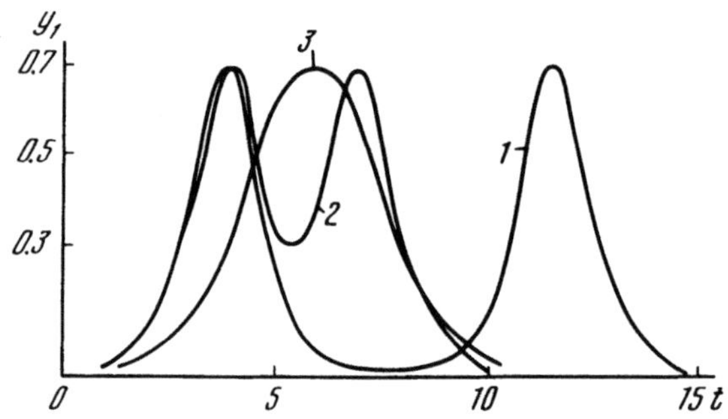

Fig. 8. The shape of the stationary pulse y_1. Curve 1) y_0 = 0.001; 2) y_0 = 0.1; 3) y_0 = 0.5.

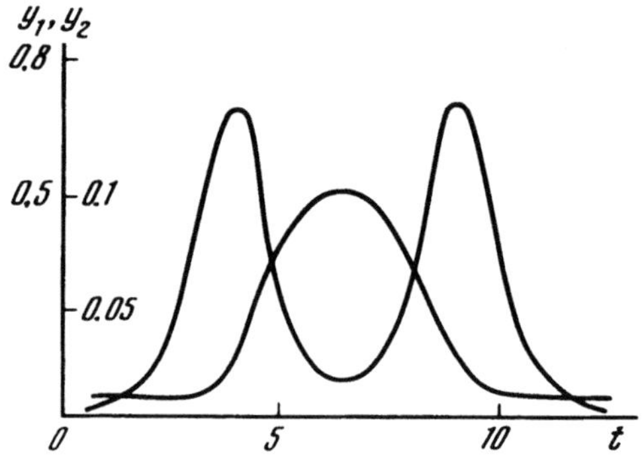

Fig. 9. The pulse shapes for $\lambda = 1$, $y_0 = 0.01$.

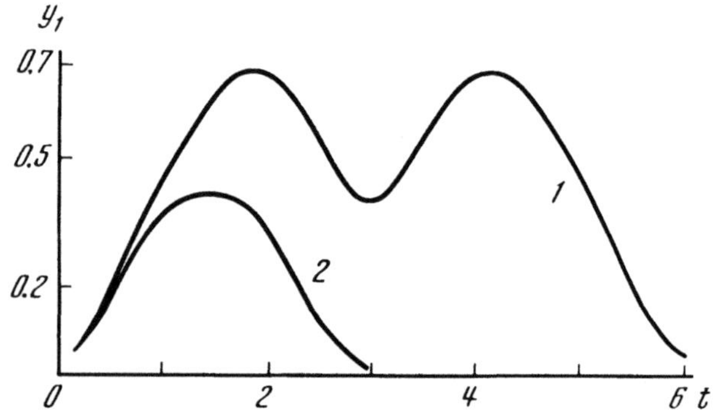

Fig. 10. Shape of the stationary pulse y_1. $C_1 = 0.6$; $C_2 = C_3 = 0$; $C_5 = 0.3$; $\lambda = 1$. Curve 1) $y_0 = 0.1$; curve 2) $y_0 = 1$.

then the pulse is double-humped; otherwise, it is single-humped. For $y_0 \to 0$ the pulse consists of the two "almost" 2π-pulses spread apart more in time the smaller the value of y_0 (Fig. 8). From Eq. (35) it is also clear that the field y_2 first rises and then falls to its initial value y_0 for an arbitrary parameter ratio. The pulse shape for $\lambda = 1$ and $y_0 = 0.01$ is shown in Fig. 9.

For arbitrary initial populations the criterion for the pulse y_1 to be double-humped is given by the expression

$$C_5 + 2C_1^2 > \frac{2}{\lambda^2}\left[C_3C_4 \sinh \lambda\pi + (C_4^2 + C_3^2)\sinh^2 \frac{\lambda\pi}{2}\right].$$

The shape of some y_1 pulses for $y_0 = 0.1$ and $y_0 = 1$ is shown in Fig. 10. The strong dependence of the shape of the y_1 pulse on the amplitude of the controlling field is obvious from the figure.

It follows from this discussion that it is possible to control the passage of ultrashort light pulses through a medium at one frequency by means of light at another frequency. By changing the amplitude of the field y_0 in the $2 \leftrightarrow 3$ transition it is possible to reduce or increase the self-transmission threshold for a pulse at the $1 \leftrightarrow 2$ transition frequency. This is especially simple to do when there is a strong difference in the dipole moments of the transitions. Then even a small change in the field y_0 may cause a sharp change in the conditions for passage of a pulse at frequency ω_{12}. In particular, a fast acting optical shutter with response time $< T_2$ and a certain steepness of the leading edge of the controlling pulse may be such a system.

Another example, illustrated by Figs. 6-10, is the possibility of modulating the shape of stationary pulses. We note that stationary self-transmission regimes are discussed above. In the nonstationary case, when the fields at the entrance to the medium differ from soliton fields, the pattern of their separation into solitons (pulse shape of the solitons, their total number, and spacing) must also depend strongly on the characteristics of the controlling field. These questions, which are of great practical interest, still await analysis.

5. Conclusion. Other Applications

It is possible to point out a number of physical applications, besides radiation control problems, which involve the approach developed here. These are, primarily, a group of problems in which collisional relaxation does not play a controlling role. They include, for example, the passage of short optical pulses through the upper atmosphere, where self-transmission effects will be important. There is also great interest in the action of radiation on a multiply charged ion plasma, where radiative processes are also dominant and collisions may be neglected.

Coherent effects may be very important in analyzing the interaction of radiation with the vapors of metals and certain other elements where the probabilities of radiative transitions are very large and a superradiant regime is observed at large pumping levels.

Generally for pulses shorter than or on the order of the collision time the population equations become inapplicable and a system of equations such as Eqs. (2), (4), (5), and (16) must be used. The formalism developed here describes the correlation among the radiating molecules caused by their coupling with the light field; i.e., it describes the kinetics of coherent relaxation. In this regard we note [8] which deals with an equidistant multilevel system in which the laser pump light can be absorbed in a transition between any pair of levels. This situation occurs if the width of the line is comparable to the frequency difference between the levels. In [8] a generalized control equation was obtained for this case, and it was shown that when coherent effects are taken into account in such a system the results are substantially different from those obtained by population kinetics (in particular, overpopulation of the upper levels). If we include broadening of the levels in the laser field, then at sufficiently high field strengths this equidistant absorption mechanism becomes dominant for multiatomic molecules at low densities when collisions may be neglected [9].

The kinetic equations discussed in this article may be supplemented by collisional terms in specific cases. This makes it possible to correctly describe relaxation in general.

LITERATURE CITED

1. T. M. Makhviladze and L. A. Shelepin, Trudy FIAN, 70:120 (1973).
2. L. A. Bol'shov and A. P. Napartovich, Zh. Éksp. Teor. Fiz., 68:1763 (1975).
3. T. M. Makhviladze and L. A. Shelepin, Zh. Éksp. Teor. Fiz., 62:2066 (1972).
4. I. M. Lizin, T. M. Makhviladze, and L. A. Shelepin, Trudy FIAN, 87:21 (1976) (present issue).
5. V. M. Fain, Quantum Electronics [in Russian], Sov. Radio, Moscow (1972), Vol. I.

6. T. M. Makhviladze, Candidate's Dissertation, Moscow Physicotechnical Institute (1972).
7. S. L. McCall and E. L. Hahn, Phys. Rev., 183:457 (1969).
8. J. Stone, E. Thiele, and M. F. Goodman, J. Chem. Phys., 59:2909, 2919 (1973).
9. V. M. Akulin, S. S. Alimpiev, N. V. Karlov, and L. A. Shelepin, Zh. Éksp. Teor. Fiz., 69:836 (1975).

DIFFERENCE METHODS AND THEIR ROLE IN THE THEORY OF COHERENT PHENOMENA

V. P. Karasev and L. A. Shelepin

The possibility of using the calculus of finite differences in the physics of coherent phenomena is examined. The dual nature of the second quantization operators is analyzed and the necessity of having a finite-difference description of coherent phenomena as well as the differential description is pointed out. A detailed discussion of finite-difference relations is given and their relationship with the generalized hypergeometric functions, the theory of orthogonal polynomials, and the Clebsch–Gordan coefficients is studied. It is shown that computational group theoretical methods and, in particular, the theory of angular momentum may be formulated in the language of finite differences. The results presented here may serve as a basis for some physical applications.

INTRODUCTION

The development of the theory of coherent cooperative phenomena requires adequate mathematical analysis. Group theoretical methods have been highly effective. The first step was made by Dicke [1] who analyzed a two-level system on the basis of the SU_2 group. Later [3-4] it was shown that coherent effects in multilevel (multiparticle) systems may be described by representations of the SU_n group (n is the number of levels).

Difference methods may play a no less important role in (the study of) these phenomena. Their importance is determined by the following two aspects. First, the analysis of the kinetics of multilevel systems using the density matrix (including cooperative effects), as in the case of population kinetics with a discrete level structure, may involve the effective utilization of finite differences. An approach based on a random walk over discrete levels (cf. [5]) may be used to discuss the kinetics of the resonant interaction of radiation and matter. If this aspect is obvious from a physical standpoint, then the second aspect, which involves the use of second quantization operators, is more subtle.

These operators are used both to construct coherent Glauber states [6] and their various generalizations [7-11], and to analyze cooperative effects in multilevel systems. In addition, the second quantization operators have a dual nature. On the one hand, they may be represented by differential operators (certain bilinear combinations of them are generators of Lie groups) and, on the other hand, they may be a special case of the generalized displacement operators [12] defined by

$$V^h f(x) = f(x+h), \tag{1}$$

where $f(x)$ is a numerical function. These operators are closely related to the calculus of finite differences. However, this relationship has not been studied sufficiently. In general, the role of difference methods is underestimated in physics as a whole, not only as applied to coherent phenomena. Thus, they are discussed rather broadly here. In Section 1 we consider the use of second-quantization operators and consider their dual nature, and in Sections 2-4 we discuss finite-difference methods and their relation to the methods of mathematical physics, in particular, groups.

1. Second-Quantization Operators and Coherent States

The role of second-quantization operators in the construction of coherent states is well known and has been discussed in a number of monographs [6, 7, 13]. However, their finite difference properties are not sufficiently used. We shall briefly discuss the role of second-quantization operators and their relationship to the coherent properties of matter.

The second-quantization operators for bosons and fermions satisfy, respectively, the relations

$$[a_i, a_k^+] = a_i a_k^+ - a_k^+ a_i = \delta_{ik}, \qquad (2)$$
$$\{b_i, b_k^+\} = b_i b_k^+ + b_k^+ b_i = \delta_{ik}.$$

The effect of the operators a_i^+, and a_i on an oscillator wave function is expressed by the equations

$$a_i^+ |n\rangle = \sqrt{n+1}\,|n+1\rangle, \qquad (3)$$
$$a_i |n\rangle = \sqrt{n}\,|n-1\rangle,$$

and that of the particle number operator $a_i^+ a_i$ by

$$a_i^+ a_i |n\rangle = n|n\rangle. \qquad (4)$$

An arbitrary state is expressed in terms of the vacuum state by

$$|n\rangle = \frac{(a^+)^n}{(n\,!)^{1/2}}|0\rangle, \qquad (5)$$
$$|n_1 n_2\rangle = \frac{(a_1^+)^{n_1}(a_2^+)^{n_2}}{(n_1\,!\,n_2\,!)^{1/2}}|00\rangle.$$

Coherent states are generally defined as the states closest to classical. Then the uncertainty relations are minimized.

Coherent states are constructed as the eigenfunctions of the annihilation operators a_i and have the form [6, 7]

$$|z\rangle = e^{-\frac{1}{2}|z|^2}\sum_{n=0}^{\infty}\frac{z^n}{(n\,!)^{1/2}}|n\rangle = e^{-\frac{1}{2}|z|^2}\sum_{n=0}^{\infty}\frac{z^n (a^+)^n}{n\,!}|0\rangle. \qquad (6)$$

This function is a superposition of states $|n\rangle$ and is expressed in terms of a polynomial series in the operator a^+.

Coherent states of the angular momentum in the form

$$|z_1 z_2\rangle = e^{-\frac{1}{2}(|z_1|+|z_2|^2)}\sum_{R=0}^{\infty}\sum_{M=-R}^{R}\frac{z_1^{R+M} z_2^{R-M}}{[(R+M)\,!\,(R-M)\,!]^{1/2}}|R, M\rangle \qquad (7)$$

have been constructed in [8]. Generalized coherent states for various systems have been constructed by a number of authors. In [14] a method was given of introducing such states for an arbitrary dynamic quantum system by constructing creation and annihilation operators from the initial coordinates and conjugate momenta of the system, after which they acquired the sense of integrals of the motion. The relation

$$K_-|z\rangle = z|z\rangle, \tag{8}$$

where K_- is the lowering operator, was used in [15] to construct generalized coherent states for the noncompact SU(1,1) group. A method was formulated in [16] for constructing generalized states for any Lie group. Although on the whole the problem of generalizing coherent states to an arbitrary system cannot be solved uniquely, the use of creation and annihilation operators lies at the basis of all cases. We note here that from the physical standpoint the correct generalization must be regarded as the principle of maximum closeness to a classical state. Then the problem of the transition to a quasiclassical description is solved by the construction of coherent states.

The group theoretical approach has also been very effective for analyzing the coherent cooperative properties of multilevel systems. In this approach an n-level system is described by representations of the SU_n group, while the quantum numbers corresponding to the signature [P, Q, R, ...] of the representations D(P, Q, R, ...) do not change for any radiative transitions between levels. The generators of the SU_n groups, the construction of representations, and the methods for calculating cooperative effects may also be described in the language of second-quantization operators. Generators of SU_n groups are constructed and written in terms of the second-quantization operators in the following. This representation of the generators in terms of fermion operators is known as the Judd representation (cf. [17, 18]).

Specific constructions are obtained on the basis of a canonical representation of the root space of Lie groups. For reference the Cartan matrices defined on the basis of the Dynkin schemes [19],

$$A_{ik} = \frac{2(\alpha_i \alpha_k)}{(\alpha_i \alpha_i)}, \tag{9}$$

are given in Appendix 1. Here α_i and α_k are the basis roots in the root space. If the first root $\alpha_1 = (\alpha_1^{(1)} \alpha_1^{(2)}, \ldots)$ is chosen to have the form $\alpha_1 = (1, 0, 0, \ldots)$, then the remaining roots α_k are easily found from the Cartan matrix. The basis roots are given in Appendix 2. For the SU_n groups the nonzero components of the basis roots α_s are $\alpha_s^{(s-1)}$ and $\alpha_s^{(s)}$, which are determined with the aid of continuous fractions. Their structure is clear from the example

$$\alpha_5^{(5)} = \sqrt{1 - \cfrac{z^2}{1 - \cfrac{z^2}{1 - \cfrac{z^2}{1 - z^2}}}}, \tag{10}$$

where $z = 1/2$. For the components $\alpha_{s+1}^{(s)}$ we have

$$\alpha_{s+1}^{(s)} = -z/\alpha_s^{(s)}. \tag{11}$$

All the basis roots may be computed using known rules [18], and by using the canonical commutation relations among the generators H_i and E_α,

$$\begin{aligned} &[H_i, H_j] = 0, \\ &[H_i, E_{\alpha_k}] = \alpha_k^i E_{\alpha_k}, \\ &[E_{\alpha_k}, E_{-\alpha_k}] = \alpha_k^i H_i, \\ &[E_\alpha, E_\beta] = N_{\alpha\beta}^\gamma E_\gamma \end{aligned} \tag{12}$$

it is possible to express the generators in terms of the operators a_i and a_i^+.

We now present the results.

The SU_2 group:

$$H_1 = \frac{1}{2}(a_1^+ a_1 - a_2^+ a_2),$$
$$E_\alpha = \frac{1}{\sqrt{2}} a_1^+ a_2, \quad E_{-\alpha} = \frac{1}{\sqrt{2}} a_2^+ a_1. \tag{13}$$

The SU_3 group:

$$H_1 = \frac{1}{2}(a_1^+ a_1 - a_2^+ a_2),$$
$$H_2 = \frac{1}{2\sqrt{3}}(a_1^+ a_1 + a_2^+ a_2 - 2a_3^+ a_3),$$
$$E_{\alpha_1} = \frac{1}{\sqrt{2}} a_1^+ a_2, \quad E_{-\alpha_1} = \frac{1}{\sqrt{2}} a_2^+ a_1, \tag{14}$$
$$E_{\alpha_2} = \frac{1}{\sqrt{2}} a_2^+ a_3, \quad E_{-\alpha_2} = \frac{1}{\sqrt{2}} a_3^+ a_2,$$
$$E_{\alpha_1+\alpha_2} = \frac{1}{\sqrt{2}} a_1^+ a_3, \quad E_{-\alpha_1-\alpha_2} = \frac{1}{\sqrt{2}} a_3^+ a_1.$$

The SU_4 group:

$$H_1 = \frac{1}{2}(a_1^+ a_1 - a_2^+ a_2),$$
$$H_2 = \frac{1}{2\sqrt{3}}(a_1^+ a_1 + a_2^+ a_2 - 2a_3^+ a_3),$$
$$H_3 = \frac{1}{2\sqrt{6}}(a_1^+ a_1 + a_2^+ a_2 + a_3^+ a_3 - 3a_4^+ a_4),$$
$$E_{\alpha_1} = \frac{1}{\sqrt{2}} a_1^+ a_2, \quad E_{-\alpha_1} = \frac{1}{\sqrt{2}} a_2^+ a_1,$$
$$E_{\alpha_2} = \frac{1}{\sqrt{2}} a_2^+ a_3, \quad E_{-\alpha_2} = \frac{1}{\sqrt{2}} a_3^+ a_2,$$
$$E_{\alpha_3} = \frac{1}{\sqrt{2}} a_3^+ a_4, \quad E_{-\alpha_3} = \frac{1}{\sqrt{2}} a_4^+ a_3, \tag{15}$$
$$E_{\alpha_1+\alpha_2} = \frac{1}{\sqrt{2}} a_1^+ a_3, \quad E_{-(\alpha_1+\alpha_2)} = \frac{1}{\sqrt{2}} a_3^+ a_1,$$
$$E_{\alpha_2+\alpha_3} = \frac{1}{\sqrt{2}} a_2^+ a_4, \quad E_{-(\alpha_2+\alpha_3)} = \frac{1}{\sqrt{2}} a_4^+ a_2,$$
$$E_{\alpha_1+\alpha_2+\alpha_3} = \frac{1}{\sqrt{2}} a_1^+ a_4, \quad E_{-(\alpha_1+\alpha_2+\alpha_3)} = \frac{1}{\sqrt{2}} a_4^+ a_1.$$

The generators of an arbitrary SU_n group are expressed in terms of the second-quantization operators in a similar manner. Then the generators of a subgroup SU_m are retained in the higher group SU_n ($m < n$). In general commutative operators for the SU_n group have the form

$$H_s = \frac{a_s^{(s)}}{s+1}\left(\sum_{i=1}^{s} a_i^+ a_i - s a_{s+1}^+ a_{s+1}\right). \tag{16}$$

Here $0 \le s \le n$ and the raising operators are

$$E_{\alpha_i} = E^{(rs)} = \frac{1}{\sqrt{2}} a_r^+ a_s,$$

where $0 \leq r \leq n$ and $r > s \geq 0$. For the lowering operators $r < s$, and the procedure for constructing the generators of other classical groups is analogous. As an example we take the $B_2(Sp4)$ group:

$$H_1 = \frac{1}{2}(a_1^+a_1 - a_2^+a_2),$$
$$H_2 = \frac{1}{2}(a_3^+a_3 - a_4^+a_4);$$
$$E_{\alpha_1} = \frac{1}{\sqrt{2}} a_1^+a_2, \qquad E_{-\alpha_1} = \frac{1}{\sqrt{2}} a_2^+a_1,$$
$$E_{\alpha_2} = \frac{1}{2}(a_2^+a_4 + a_3^+a_1), \qquad E_{-\alpha_2} = \frac{1}{2}(a_4^+a_2 + a_1^+a_3), \tag{17}$$
$$E_{\alpha_1+\alpha_2} = \frac{1}{2}(a_1^+a_4 - a_3^+a_2), \qquad E_{-\alpha_1-\alpha_2} = \frac{1}{2}(a_4^+a_1 - a_2^+a_3),$$
$$E_{\alpha_1+2\alpha_2} = \frac{1}{\sqrt{2}} a_3^+a_4, \qquad E_{-\alpha_1-2\alpha_2} = \frac{1}{\sqrt{2}} a_4^+a_3.$$

We note that the second-quantization operators have tensor character and are classified according to representations of the corresponding group. For example, the operators (14) transform according to the representation D(11), and the operators a_i and a_i^+, themselves, according to the representations D(10) and (01).

We note that the coherent properties of systems with a fixed number of particles of different kinds were described in [20]. The coherent characteristics of such systems were derived from the tensor properties of the second-quantization operators.

The generators of a Lie group may be expressed in terms of bilinear combinations of fermion operators as well as of boson operators. As an example we give the generators of the SU_2 groups written with tensor operators [17, 18]:

$$H_1 = -\frac{1}{4}\{[S][l]\}^{1/2}\{(a^+a)^{00} + (aa^+)^{00}\},$$
$$E_\alpha = \frac{1}{2}\{[S][l]\}^{1/2}(a^+a^+)^{00}, \tag{18}$$
$$E_{-\alpha} = -\frac{1}{2}\{[S][l]\}^{1/2}(aa)^{00},$$

where $[S] = 2S + 1$ and the indices 00 denote coupling of the tensors a^+ and a by means of the Clebsch–Gordan coefficients for the spin S and orbital angular momentum l. The generators of any Lie group may be expressed analogously in terms of the (boson or fermion) second-quantization operators.

Furthermore, all states of a system which correspond to the basis vectors of irreducible representations can be obtained by operating with polynomials in the generators on a function corresponding to the leading term (weight). Thus, polynomials made up of second-quantization operators may indeed play a universal role in group theoretical analyses. They are used to write the projection operators which play a fundamental role in the evaluation of various effects [21-24] and in the calculation of the Clebsch–Gordan and Casimir coefficients which are used to construct the invariant (conserved) quantities.

With these polynomials it is possible to construct an algebra which corresponds to generalized Clebsch–Gordan coefficients (cf. [25, 26]) and specifies the coefficients in the invariant equations (of the type found in a Dirac algebra), and to construct finite transformation operators which are of great significance in evaluating coherent effects. Without going into specifics on each of these possibilities, we only emphasize that it is possible to treat practically all applied group theory problems, including the analysis of coherent effects, in the language of polynomials in the second-quantization operators.

In addition, the second-quantization operators have a two-in-one character. They have dual properties. On the one hand, they are isomorphic with differential operators. Thus, for the SU_2 group we have

$$\begin{gathered} a_k = \frac{1}{\sqrt{2}}(p_k + iq_k), \quad a_k^+ = \frac{1}{\sqrt{2}}(p_k - iq_k), \\ p = -i\frac{\partial}{\partial x^k}, \quad q_k = x_k \quad (k = 1, 2, 3) \\ J_k = i\varepsilon_{klm} a_l^+ a_m. \end{gathered} \tag{19}$$

For the SU_n group we have

$$\begin{gathered} A_k^i(x) = x_k \frac{\partial}{\partial x_i} - \frac{\delta_k^i}{n} \sum_{j=1}^{n} x_j \frac{\partial}{\partial x_j}, \\ A_k^i(\xi) = -\xi^i \frac{\partial}{\partial \xi^k} + \frac{\delta_k^i}{n} \sum_{j=1}^{n} \xi^j \frac{\partial}{\partial \xi^j}, \end{gathered} \tag{20}$$

where $E_{-\alpha} = A_k^i$ $(i < k, \alpha > 0)$, $E_\alpha = A_k^i$ $(i > k, \alpha > 0)$, and $H_i = A_i^i - A_{i+1}^{i+1}$, $\xi = x^*$. However, there is another side. The finite-difference differentiation operator, defined by

$$\Delta_n f = f(n+1) - f(n), \tag{21}$$

is directly expressed in terms of the second-quantization operators by

$$\Delta_n = \frac{a^+}{[(n+1)!]^{1/2}} - 1 \quad \left(\text{or } \Delta_{-n} = 1 - \frac{a}{(n!)^{1/2}}\right). \tag{22}$$

The differential properties of the generators (and second-quantization operators) as applied to computational group theoretical methods have been used fairly widely. It should be noted that in the analysis of cooperative effects the phenomenological description based on nonlinear differential equations is important in a number of cases. This approach was originated by Ginzburg and Landau [26], who proposed a nonlinear system of equations for superconductivity. It is now used for other coherent phenomena as well. However, the relation between the phenomenological nonlinear equations and the group theoretical approach has still not been adequately studied.

Proceeding from the above we might expect that finite differences would be an effective instrument for describing coherent effects. However, despite the available fundamental possibilities, the difference method is practically unused. The aim of the following sections is to study the relationship between the finite difference and group methods and to develop the basis for their practical utilization.

2. Some Problems in Discrete Analysis (the Calculus of Finite Differences)

The theory of finite differences found physical applications in the early stages of its development since the traditional research problems in this theory, the interpolation of functions, was closely related to the treatment of data [27-29]. Recently, however, the range of physical applications of the theory of finite differences has expanded substantially. This is primarily due to the use of difference methods to solve the boundary value problems of mathematical physics [30-31] by approximating the differential equations with difference equations and then using an electronic computer for the calculations. In [32] difference methods were used to analyze and obtain an exact solution of the relativistic Coulomb problem within the framework of a quasipotential approach. We have noted above the prospects for using finite differences in the theory of coherent cooperative phenomena. There is special interest in the use of the calculus of finite differences in a new (recently formulated) branch of mathematical physics, the theory of angular momenta and its generalizations [33-34]; this is for two reasons. First, the range of physical applications of the "symmetry formalism" (the theory of angular momenta and its applications) is constantly expanding (see, for example, Section 1 of this paper and [34]); and, second, the applications of this formalism lead to new problems and to new methods of analyzing and solving the traditional problems of finite-difference theory.

In this section we present the basic concepts and definitions of discrete analysis (the theory of finite differences) and develop the special problems of importance both for the direct development of the theory of coherent cooperative phenomena and for the development of the symmetry formalism. The results obtained here will then be used in the succeeding sections of this paper. The contents of all these sections (2-4) are mainly based on material in the unpublished work "Selected Problems in Discrete Analysis and Its Applications" by V. P. Karasev (May 1975), and on [35].

2.1. Elements of the Theory of Finite Differences (General Points)

The central place in the calculus of finite differences is occupied by the linear operators Δ_x and $P_{(x)}^{-1}$ which play roles analogous to differentiation, $D_x \equiv d/dx$, and the definite integral $\int_0^x dt$ in continuous analysis [27-30]. These operators are defined by

$$\Delta_x f(x) = (V_x - \mathbf{1}) f(x) \equiv f(x+1) - f(x), \tag{23a}$$

$$P_{(x)}^{-1}\{f(t)\} = \sum_{t=0}^{x-1} V_\tau^t f(0) \equiv \sum_{t=0}^{x-1} f(t), \tag{23b}$$

where $\mathbf{1}$ is the unit operator; V_x is the operator for displacement of the argument x of a function by unity; and, $V_\tau^t f(0) = V_\tau V_\tau^{t-1} f(0) = V_\tau^t f(\tau)|_{\tau=0}$.

With these definitions it is possible to express other operators of this calculus [27-29] in terms of Δ_x and $P_{(x)}^{-1}$, and also to obtain a number of useful formulas. As examples we shall write a formula for the s-th (s is a natural number) iteration of the operator Δ_x:

$$\Delta_x^s f(x) = (V_x - 1)^s f(x) = \sum_{\sigma=0}^{s} \frac{(-1)^{s+\sigma} s!}{\sigma!(s-\sigma)!} V_x^\sigma f(x) = \sum_{\sigma=0}^{s} \frac{(-1)^{s+\sigma} s!}{\sigma!(s-\sigma)!} f(x+\sigma), \tag{24}$$

and a difference analog of the Newton–Leibniz formula of integral calculus:

$$P_{(x)}^{-1}\{\Delta_t f(t)\} \equiv \sum_{t=0}^{x-1} \Delta_t f(t) = f(x) - f(0). \tag{25}$$

There are still a number of discrete (difference) analogs for the results (formulas, theorems) of continuous analysis. Thus, for example, the difference analog of the Leibniz formula of differential calculus has the form

$$\Delta_x^s [u(x) v(x)] = \sum_{n,k} \frac{s!\, \Delta_x^k u(x)\, \Delta_x^n v(x)}{(n+k-s)!(s-k)!(s-n)!}. \tag{26}$$

The limits of the sum here and in the following (when they are not indicated) are determined by the convention that the reciprocals of the factorials of negative numbers be taken equal to zero [29].

The difference analog of the formula for integrating by parts is Abel's indentity,

$$P^{-1}_{(x)}\{u(t)\Delta_t v(t)\} \equiv \sum_{t=0}^{x-1} u(t)\Delta_t v(t) = u(x)v(x) - u(0)v(0) - \sum_{t=0}^{x-1} v(t+1)\Delta_t u(t). \tag{27}$$

There are a number of functions for which the evaluation of $\Delta^s_x f(x)$ and $P^{-1}_{(x)}\{f(t)\}$ is determined by simple equations [27-29]. The most important of these are the exponential (power) function a^x and the symbolic (difference) power (or factorial function) $x^{(n)}$ which is defined for natural x and n by the formulas

$$x^{(n)} \equiv \frac{x!}{(x-n)!} = x(x-1)\dots(x-n+1),$$
$$x^{(-n)} \equiv \frac{x!}{(x+n)!}. \tag{28}$$

For general x and n the factorials in this definition are replaced by gamma functions, with $x! \to \Gamma(x+1)$, etc. We note that for natural n, $x^{(n)}$ is a polynomial in x (the Bernoulli polynomials $B^{(n+1)}_n(x)$ [28]), where the coefficients of x^k are the Stirling numbers of the first kind s(n, k) (see below).

The results of difference "differentiation" and "integration" of these functions are given by the following equations:

$$\Delta_x x^{(n)} = nx^{(n-1)}, \quad \Delta^s_x x^{(n)} = \frac{n!}{(n-s)!} x^{(n-s)}; \tag{29a}$$

$$\Delta_x a^x = (a-1)a^x, \quad \Delta^s_x a^x = (a-1)^s a^x, \tag{29b}$$

$$P^{-1}_{(x)}\{t^{(n)}\} \equiv \sum_{t=0}^{x-1} t^{(n)} = \frac{x^{(n+1)} - 0^{(n+1)}}{n+1} = \frac{x^{(n+1)}}{(n+1)}, \tag{29c}$$

$$P^{-1}_{(x)}\{a^t\} \equiv \sum_{t=0}^{x-1} a^t = \frac{a^x-1}{a-1}. \tag{29d}$$

It is easy to establish a number of relations for the symbolic powers:

$$(-a)^{(n)} = (-1)^n/(a-1)^{(-n)} = (-1)^n (a-1+n)^{(n)}; \tag{30a}$$

$$a^{(m+k)} = a^{(m)}(a-m)^{(k)}; \tag{30b}$$

$$(a+k)^{(\alpha)} = \frac{a^{(\alpha)}(a+k)^{(k)}}{(a-\alpha+k)^{(k)}} = \frac{a^{(\alpha)}(-a-1)^{(k)}}{(-a+\alpha-1)^{(k)}}, \tag{30c}$$

$$(a-k)^{(\alpha)} = a^{(\alpha)}(a-\alpha)^{(k)}/a^{(k)}, \tag{30d}$$

which together with properties (29a) and (29c) allow them to play a role in discrete analysis analogous to that of the ordinary power in continuous analysis. In particular, it is possible to expand a function in a series of symbolic powers (factorial series [28]) analogous to the usual power series:

$$f(x) = \sum_{n=0}^{\infty} a_n (x-l)^{(n)}, \quad a_n = \frac{1}{n!}\Delta^n_x f(x)|_{x=l}, \tag{31a}$$

$$F(x) = \sum_{k=0}^{\infty} b_k (-1)^k (x-l)^{(-k-1)} k!. \tag{31b}$$

Thus, for the functions $(a+x)^{(n)}$, $(b-x)^{(m)}$, and $a^{\alpha x}$, we can, with Eqs. (29) and (30), easily establish the expansions

$$(a+x)^{(n)}=\sum_k \frac{n!}{k!(n-k)!}x^{(k)}a^{(n-k)}, \tag{32a}$$

$$(b-x)^{(m)}=(-1)^m(x-b+m-1)^{(m)}=\sum_k \frac{m!(-1)^k x^{(k)}(b-k)^{(m-k)}}{k!(m-k)!}, \tag{32b}$$

$$a^{\alpha x}=\sum_k \frac{x^{(k)}}{k!}(a^\alpha-1)^k, \tag{32c}$$

which are analogous to the usual binomial (32a, 32b) and exponential (32c) series. Other expansions of this type are given below during the analysis of specific problems.

These examples indicate the existence of analogies between continuous and discrete analysis which allow us to carry over results from the former to the latter in a number of cases by making the formal substitutions

$$x^k \div\!\!\to x^{(k)}; \tag{33a}$$

$$D_x^s \div\!\!\to \Delta_x^s \tag{33b}$$

in the equations and formulas of continuous analysis. We shall call this substitution "discretization." (In the following we shall denote the results of applying Eq. (33) by the sign ~ placed on top and write the variables being made discrete on the lower right.)

There is, however, not a complete isomorphism between discrete and continuous analysis. This is reflected both in the explicit form of the existing analogies [see, for example, Eqs. (26) and (27)] and in the impossibility of discretizing certain results of continuous analysis. Thus, there is no general difference analog of the formula for differentiating a composite function. That is,

$$\Delta_x f(u(x)) \neq \Delta_u f(u)\,\Delta_x u(x)\,|_{u=u(x)} \tag{34}$$

for an arbitrary function u(x); however, for $u_1(x) = a + x$ and $u_2(x) = a - x$ (a constant) we have

$$\Delta_x f(a+x)=\Delta_{a+x}f(a+x)=\Delta_u f(u)\,|_{u=a+x}\equiv\Delta_u f(u)\,\Delta_x u(x)\,|_{u=a+x}, \tag{35a}$$

$$\Delta_x^s f(a-x)=(-1)^s\Delta_u^s f(u-s)\,|_{u=a-x}\equiv\Delta_u^s f(u-s)\,\Delta_x^s u(x)\,|_{u=a-x}. \tag{35b}$$

In addition, as opposed to the simple law for multiplying ordinary powers, $x^n x^m = x^{n+m}$, the rule for multiplying symbolic powers is more complicated:

$$x^{(m)}x^{(n)}=\sum_l C_l^{m,n}x^{(l)},$$

$$C_l^{m,n}\equiv\frac{1}{l!}\Delta_x^l[x^{(n)}x^{(m)}]|_{x=0}=\frac{m!\,n!}{(l-m)!\,(l-n)!\,(m+n-l)!}. \tag{36}$$

In the case of n (n is any natural number) factors the result is still more complicated, but symmetric:

$$\prod_{i=1}^n x^{(r_i)}=\sum_{l\geqslant\max\{r_i\}}^R \frac{x^{(l)}}{l!}\,{}_nF_{n-1}\left(\begin{matrix}(r_i-l)_n\\(-l)_{n-1}\end{matrix};1\right)\prod_{i=1}^n l^{(r_i)}, \tag{37}$$

where $R=\sum_{i=1}^n r_i$ and ${}_nF_{n-1}\left(\begin{matrix}\cdots\\\cdots\end{matrix};1\right)$ is a generalized hypergeometric series. (The derivation of this formula is based on the results of Section 2.2, to which we also refer for the notation employed here.) All this complicates the transfer of results and methods from continuous analysis to discrete analysis, in particular, to the practically important (see Sections 3 and 4) theory of difference equations (the discrete analog of the theory of differential equations). For example, the direct difference analog of the simple relation $x^r D_x^k x^n = [n!/(n-k)!]x^{n-k+r}$ has the form [in accordance with Eq. (33)]

$$x^{(r)}\Delta_x^k x^{(n)} = \sum_l \frac{x^{(l)} r!\, n!}{(r+n-k-l)!\,(l-r)!\,(l-n+k)!}, \tag{38}$$

which makes solution of difference equations with the aid of series expansions in $x^{(n)}$ difficult. In this regard, the most widely used techniques for solving difference equations are machine methods [30], which are satisfactory for calculations but have limited possibilities for qualitative analysis of the solutions.

Of the analytic methods used in various branches of discrete analysis, some symbolic (operational) procedures [28] are fruitful. They primarily rely on certain (defined by formal power series) functions of the operators Δ_x and V_x and an operator algebra [36], as well as the simplicity of the expressions for $\Delta_x^k f(x)$ and $V_x^n f(x)$ [see Eqs. (23) and (24)].

These methods make it possible to carry out complicated transformations and to easily obtain certain results which are exceedingly difficult to obtain by conventional means [28]. We note, however, that as in the case of continuous analysis the results obtained in this way must be justified. Nevertheless, for some classes of functions $f(x)$ (for example, the polynomials $f(x) = \sum_n a_n x^{(n)}$) operational methods may be used automatically. Thus, in the following they will often be applied either to obtain the final results or to advance reasonable hypotheses. As examples we shall write a number of formal operator relations which will be used later:

$$\Delta_x = V_x - 1, \quad V_x = \Delta_x + 1; \tag{39a}$$

$$V_x^{-1} = \frac{1}{1+\Delta_x} = \sum_{k=0}^{\infty} (-1)^k \Delta_x^k; \tag{39b}$$

$$\ln(1+\Delta_x) = \sum_{k=0}^{\infty} \frac{(-1)^k \Delta_x^{k+1}}{k+1} = -\sum_{k=1}^{\infty} \frac{(-1)^k \Delta_x^k}{k}; \tag{39c}$$

$$\Delta_x^\alpha = (-1)^\alpha_1 F_0(^{-\alpha};\, V_x) \equiv \sum_{k=0}^{\infty} \frac{\alpha^{(k)}}{k!} (-1)^{\alpha+k} V_x^\alpha, \tag{39d}$$

where α is an arbitrary real number and ${}_1F_0(\cdots;\ \ldots)$ is the generalized hypergeometric series. In addition we shall now illustrate the operational method by deriving two useful formulas with it.

Applying Eq. (39d) with α = s (s a natural number) to $\prod_{i=1}^{n} u_i(x)$ and including the fact that

$$V_x^k \prod_{i=1}^{n} u_i(x) = \prod_{i=1}^{n} (V_x^k u_i(x)) = V_x^k u_1(x)\, V_x^k u_2(x) \ldots V_x^k u_n(x), \tag{40}$$

we find that

$$\Delta_x^s \left[\prod_{i=1}^{n} u_i(x)\right] = \sum_k \frac{s^{(k)}}{k!} (-1)^{s+k} \prod_{i=1}^{n} (V_x^k u_i(x)). \tag{41}$$

Replacing V_x by $\Delta_x + 1$ in Eq. (41), carrying out binomial expansion $(\Delta_x + 1)^k$, and changing the order of summation, we find the final result,

$$\Delta_x^s\left[\prod_{i=1}^{n} u_i(x)\right] = \sum_{\{\alpha_i\}} {}_nF_{n-1}\left({(\alpha_i - s)_n; \atop (-s)_{n-1}} 1\right) \prod_{i=1}^{n} \frac{s^{(\alpha_i)}}{\alpha_i!} \Delta_x^{\alpha_i} u_i(x) = \sum_{\{\alpha_i\}} {}_nF_{n-1}\left({(\alpha_i - s)_n; \atop (-s)_{n-1}} 1\right) \prod_{i=1}^{n} {}_2F_1\left({-\alpha_i,\ s-\alpha_i; \atop -\alpha_i,} 1\right) \Delta_x^{\alpha_i} u_i(x), \tag{42}$$

which is a generalization of Eq. (26) to the case of an arbitrary number of factors under the Δ_x^s sign.

The other result has to do with the symbol $\Delta_x^s \Delta_y^{x+a}$. Using Eqs. (24) and (39d) together with $\Delta_x^s[(a+x)^{(\beta)}(-1)^x] = (-1)^{s+x}(a+x)^{(\beta)}\,{}_2F_1\left({-s,\ a+x+1 \atop a+x-\beta+1}; -1\right)$ (see Section 2.2), we easily establish the following useful formula:

$$\begin{aligned}\Delta_x^s \Delta_y^{x+a} f(y) &= \sum_l \frac{(-1)^{l+a}}{l!} f(y+l)\,\Delta_x^s[(a+x)^{(l)}(-1)^x] \\ &= (-1)^{a+x+s} \sum_l \frac{(-1)^l}{l!}(a+x)^{(l)}\,{}_2F_1\left({-s,\ a+x+1 \atop a+x-l+1}; -1\right) V_y^l f(y).\end{aligned} \tag{43}$$

The appearance of the generalized hypergeometric functions ${}_pF_q(\because; 1)$ in Eqs. (37), (42), and (43) leads us to conclude that these functions are important in the development of discrete analysis. Thus, we conclude this introduction to the foundations of discrete analysis and proceed to develop specialized topics beginning with a discussion of the hypergeometric functions from the standpoint of the calculus of finite differences. In doing this the guiding idea will be to establish the interrelationships and to use the analogs between discrete and continuous analysis.

2.2. Hypergeometric Functions from the Standpoint of Discrete Analysis

The generalized hypergeometric functions, ${}_pF_q(: : :;\ z)$, are defined by series [37, 38] of the form

$${}_pF_q\left({(a_i)_p \atop (b_j)_q}; z\right) \equiv \sum_{n=0}^{\infty} \alpha_n \frac{z^n}{n!} = \sum_{n=0}^{\infty} \prod_{i=1}^{p} \frac{(a_i - 1 + n)!}{(a_i - 1)!} \prod_{j=1}^{q} \frac{(b_j - 1)!}{(b_j - 1 + n)!} \frac{z^n}{n!}, \tag{44}$$

Here and in the following the symbol $(a_i)_p$ denotes the numerical sequence $(a_1, \ldots, a_p)$ of length p; a_i and b_j are called the parameters, and z is the argument of the series. The series of equations (44) are evidently invariant with respect to interchanges of the parameters a_i and b_j in each line individually. In the particular case p = 2 and q = 1 we obtain the Gaussian series which defines the ordinary hypergeometric function and is used to express many special functions of mathematical physics. Recalling Eq. (28), the definition (44) of the hypergeometric functions can be rewritten in terms of symbolic (difference) powers as

$$\begin{aligned}{}_pF_q\left({(a_i)_p \atop (b_j)_q}; z\right) &= \sum_{n=0}^{\infty} \prod_{i=1}^{p} (a_i - 1 + n)^{(n)} \prod_{j=1}^{q} \frac{1}{(b_j - 1 + n)^{(n)}} \frac{z^n}{n!} \\ &= \sum_{n=0}^{\infty} \prod_{i=1}^{p} (-a_i)^{(n)} \prod_{j=1}^{q} \frac{1}{(-b_j)^{(n)}} (-1)^{np-nq} \frac{z^n}{n!}.\end{aligned} \tag{45}$$

This expression contains both ordinary and difference powers and may be treated as a power series of mixed type. In this regard, we shall rerer to the parameters a_i and b_j as "discrete" and the argument x as "continuous."

The functions ${}_pF_q\left(\begin{matrix}(a_i)_p\\(b_j)_q\end{matrix}; z\right)$ (${}_pF_q$ functions) satisfy the differential equation

$$\left[xD_x\prod_{j=1}^{q}(xD_x+b_j-1)-x\prod_{i=1}^{p}(xD_x+a_i)\right]{}_pF_q\left(\begin{matrix}(a_i)_p\\(b_j)_q\end{matrix}; x\right)=0, \tag{46}$$

which confirms their adherence to continuous analysis. In addition, they satisfy linear recurrence relations which relate the ${}_pF_q$ functions to those differing by ± 1 in the value of one of the parameters (while the others remain fixed) [37]; this makes them an object of discrete analysis.

The series (44) terminate and become polynomials if beginning with some number $n = N$ all the coefficients α_n ($n \geq N$) go to zero. Together with the requirement that the nonzero coefficients α_n be finite (convergence condition), this results in the following conditions for termination of the series of Eq. (44):

(a) The number of integral nonpositive $a_i \equiv a_i^+$ must exceed by at least unity the number of such $b_j \equiv b_j^+$,

(b) When a_i^+ and b_j^+ are ordered as $a_1^+ \geqslant a_2^+ \geqslant \ldots \geqslant a_m^+ \geqslant a_{m+1}^+ \geqslant \ldots \geqslant a_{m+k}^+$, and $b_1^+ \geqslant b_2^+ \geqslant \ldots \geqslant b_m^+$, the inequalities $a_i^+ \geq b_i^+$, where $i = \overline{1, m}$, must be satisfied. (Here and in the following the notation $i = \overline{1, m}$ means that the index i runs through the values $i = 1, 2, \ldots, m$.)

A very interesting fact about the ${}_pF_q$ functions is that the operation of Eq. (33) (replacing the ordinary powers z^n by the symbolic powers $z^{(n)}$) on ${}_pF_q(\because; z)$ does not change its being a ${}_pF_q$ function. The difference analogs ${}_p\tilde{F}_q(\because; x)_x$ of the ordinary ${}_pF_q$ functions are also ${}_pF_q$ functions but with changed numbers of parameters and values of the arguments [35]. We note that this "discretization" does not always yield converging series. However, as a rule, such diverging series may be interpreted with the aid of a generalization of the ${}_pF_q$ functions, the G function of Meijer, and the E function of MacRobert [32, 37]. We now illustrate these points with some examples:

$${}_p\tilde{F}_q\left(\begin{matrix}(a_i)_p\\(b_j)_q\end{matrix}; x\right)_x=\sum_{n=0}^{\infty}\prod_{i=1}^{p}(-a_i)^{(n)}\prod_{j=1}^{q}\frac{1}{(-b_j)^{(n)}}x^{(n)}\frac{(-1)^{np-nq}}{n!}={}_{p+1}F_q\left(\begin{matrix}(a_i)_p,\ -x\\(b_j)_q\end{matrix}; -1\right), \tag{47a}$$

$${}_p\tilde{F}_q\left(\begin{matrix}(a_i)_p\\(b_j)_q\end{matrix}; \frac{x}{y}\right)_{\frac{x}{y}}={}_{p+1}F_q\left(\begin{matrix}(a_i)_p,\ -\frac{x}{y}\\(b_j)_q\end{matrix}; -1\right), \tag{47b}$$

$${}_p\tilde{F}_q\left(\begin{matrix}(a_i)_p\\(b_j)_q\end{matrix}; \frac{x}{y}\right)_x={}_{p+1}F_q\left(\begin{matrix}(a_i)_p\ -x\\(b_j)_q\end{matrix}; -\frac{1}{y}\right), \tag{47c}$$

$${}_p\tilde{F}_q\left(\begin{matrix}(a_i)_p\\(b_j)_q\end{matrix}; \frac{x}{y}\right)_{x,y}={}_{p+1}F_{q+1}\left(\begin{matrix}(a_i)_p,\ -x\\(b_j)_q,\ -y\end{matrix}; 1\right), \tag{47d}$$

$${}_p\tilde{F}_q\left(\begin{matrix}(a_i)_p\\(b_j)_q\end{matrix}; -xy\right)_x={}_{p+1}F_q\left(\begin{matrix}(a_i)_p,\ -x\\(b_j)_q\end{matrix}; y\right). \tag{47e}$$

From the standpoint of applications in the development of the symmetry formalism there is particular interest in the appearance among the quantities ${}_p\tilde{F}_q(\because; x)_x$ of ${}_mF_n$ functions with a "continuous" argument equal to 1, since these are used to express a number of the basic quantities in the theory of angular momenta (see the following sections). This fact is an indication of the existence of a close interrelationship and of a unique symmetry between the discrete and continuous in the domain of the hypergeometric functions. It may find application during the transformation of results [in the sense of the operations of Eq. (33) and their "inverses"] from continuous analysis to discrete analysis [for example, when constructing difference analogs of Eqs. (46)] and vice versa.

Another remarkable fact established in [35] for finite (terminating) hypergeometric series is the ability to represent them compactly with constructions consisting of the operators Δ_x^s (s a natural number) applied to arguments made up of symbolic powers. This representation (for brevity we shall call it the CR representation) recalls the form of the generalized Rodriguez formula [37] for classical orthogonal polynominals,

$$p_n(x) = \frac{1}{K_n w(x)} D_x^n [w(x) X^n(x)] \tag{48}$$

[w(x) is a weight function; X(x) is a polynomial with coefficients independent of n; and K_n is a constant) and, in fact, extends its difference analog to arbitrary ${}_pF_q$ functions. The method of obtaining the CR representation is based on the fact that the result of applying Δ_x^s to monomials $\Psi(x)$, made up of the product of symbolic powers and power functions, is the original function $\Psi(x)$ (generating function) multiplied by some ${}_pF_q$ function. We now illustrate this by listing the results of applying Δ_x^s to some basic "elementary" generating functions $\Psi_i^0(x)$:

$$\Delta_x^s (a-x)^{(\alpha)} = (-1)^s \frac{\alpha^{(s)} (a-x)^{(\alpha)}}{(a-x)^{(s)}} = \sum_{k=0}^{s} \frac{(-1)^{k+s} s!}{k!(s-k)!} (a-x-k)^{(\alpha)} =$$

$$= \sum_{k=0}^{\infty} \frac{(a-x)^{(\alpha)} (a-x-\alpha)^{(k)} s^{(k)}}{(a-x)^{(k)}} \frac{(-1)^{k+s}}{k!} \equiv (-1)^s (a-x)^{(\alpha)} {}_2F_1\left(\begin{matrix} x+\alpha-a, & -s \\ x-a \end{matrix}; 1\right), \tag{49a}$$

$$\Delta_x^s (a+x)^{(\alpha)} = \alpha^{(s)} (a+x)^{(\alpha-s)} = (-1)^s (a+x)^{(\alpha)} {}_2F_1\left(\begin{matrix} a+x+1, & -s \\ a+x-\alpha+1 \end{matrix}; 1\right), \tag{49b}$$

$$\Delta_x^s [(a+x)^{(\alpha)}]^{-1} = \frac{(-\alpha)^{(s)}}{(a+x-s)^{(\alpha+s)}} = (-1)^s [(a+x)^{(\alpha)}]^{-1} {}_2F_1\left(\begin{matrix} a+x-\alpha+1, & -s \\ a+x+1 \end{matrix}; 1\right), \tag{49c}$$

$$\Delta_x^s a^{(x)} = (-1)^s a^{(x)} {}_2F_0\left(\begin{matrix} x-a, & -s \\ & \end{matrix}; -1\right), \tag{49d}$$

$$\Delta_x^s (b+x)^{(a+x)} = (-1)^s (b+x)^{(a+x)} {}_2F_0\left(\begin{matrix} b+x+1, & -s \\ & \end{matrix}; 1\right), \tag{49e}$$

$$\Delta_x^s (b-x)^{(a-x)} = (-1)^s (b-x)^{(a-x)} {}_1F_1\left(\begin{matrix} -s \\ x-b \end{matrix}; -1\right), \tag{49f}$$

$$\Delta_x^s a^x = a^x (a-1)^s = (-1)^s a^x {}_1F_0\left(\begin{matrix} -s \\ \end{matrix}; a\right), \tag{49g}$$

$$\Delta_x^s [(a-x)^{(\alpha)}]^{-1} = (-1)^s [(a-x)^{(\alpha)}]^{-1} {}_2F_1\left(\begin{matrix} x-a, & -s \\ x+\alpha-a \end{matrix}; 1\right), \tag{49h}$$

$$\Delta_x^s [a^{(\alpha-x)}] = (-1)^s a^{(\alpha-x)} {}_1F_1\left(\begin{matrix} -s \\ a-\alpha+x+1 \end{matrix}; -1\right), \tag{49i}$$

$$\Delta_x^s [(b+x)^{(a+x)}]^{-1} = (-1)^s [(b+x)^{(a+x)}]^{-1} {}_1F_1\left(\begin{matrix} -s \\ b+x+1 \end{matrix}; -1\right), \tag{49j}$$

$$\Delta_x^s [(a+x)^{(b+2x)}] = (-1)^s (a+x)^{(b+2x)} {}_3F_0\left(\begin{matrix} b+x-a, & a+x+1, & -s \\ & & \end{matrix}; -1\right), \tag{49k}$$

$$\Delta_x^s [(a+x)^{(b+2x)}]^{-1} = (-1)^s [(a+x)^{(b+2x)}]^{-1} {}_1F_2\left(\begin{matrix} -s \\ b+x-a, & a+x+1 \end{matrix}; -1\right). \tag{49l}$$

These equations may be easily derived using Eqs. (24), (29), (30), and (44). An example of such a derivation is shown explicitly in Eq. (49a). All these series are evidently finite. Dividing $\Delta_x^s \Psi_i^0(x)$ in Eqs. (49) by $(-1)^s \Psi_i^0(x)$ we obtain the CR representations for the ${}_pF_q$ functions on the right sides of these equations.

The CR representations for more general ${}_pF_q$ functions are obtained by taking various products of "elementary" generating functions for $\Psi(x)$. Here are some examples:

$${}_2F_1\left({x-a,\ -s \atop x-b};\ z\right)=(-1)^s\frac{b^{(x)}}{a^{(x)}z^x}\Delta_x^s\left[\frac{a^{(x)}z^x}{b^{(x)}}\right], \tag{50a}$$

$${}_2F_2\left({-s,\ a+x+1 \atop a+x-\alpha+1,\ x-b};\ -z\right)=[z^x(a+x)^{(\alpha)}/b^{(x)}]^{-1}(-1)^s\Delta_x^s[z^x(a+x)^{\alpha}/b^{(x)}], \tag{50b}$$

$${}_3F_1\left({-s,\ a+x+1,\ x-b \atop a+x-\alpha+1};\ z\right)=[(-z)^x(a+x)^{(\alpha)}b^{(x)}]^{-1}(-1)^s\Delta_x^s[(-z)^x(a+x)^{(\alpha)}b^{(x)}], \tag{50c}$$

$${}_3F_2\left({a+x+1,\ x-b+\beta,\ -s \atop x-b,\ a+x-\alpha+1};\ 1\right)=[(a+x)^{(\alpha)}(b-x)^{\beta}]^{-1}(-1)^s\Delta_x^s[(a+x)^{(\alpha)}(b-x)^{(\beta)}], \tag{50d}$$

$${}_3F_2\left({x-b,\ a+x-\alpha+1,\ -s \atop x-b+\beta,\ a+x+1};\ 1\right)=(a+x)^{(\alpha)}(b-x)^{(\beta)}(-1)^s\Delta_x^s[(a+x)^{(\alpha)}(b-x)^{(\beta)}]^{-1}; \tag{50e}$$

$${}_3F_2\left({a+x+1,\ b+x-\beta+1,\ -s \atop a+x+1-\alpha,\ b+x+1};\ 1\right)=(-1)^s\frac{(b+x)^{(\beta)}}{(a+x)^{(\alpha)}}\Delta_x^s\left[\frac{(a+x)^{(\alpha)}}{(b+x)^{(\beta)}}\right], \tag{50f}$$

$${}_3F_0\left(b+x-a,\ a+x+1,\ -s;\ z\right)=(-1)^s[(a+x)^{(b+2x)}(-z)^x]^{-1}\Delta_x^s[(a+x)^{(b+2x)}(-z)^x]; \tag{50g}$$

$${}_5F_4\left({a+x+1,\ x-b+\beta,\ x-d,\ x+c-\gamma+1,\ -s \atop a+x-\alpha+1,\ x-b,\ x-d+\varepsilon,\ x+c+1};\ 1\right)=(-1)^s\left[\frac{(c+x)^{(\gamma)}(d-x)^{(\varepsilon)}}{(a+x)^{(\alpha)}(b-x)^{(\beta)}}\right]\Delta_x^s\left[\frac{(a+x)^{(\alpha)}(b-x)^{(\beta)}}{(c+x)^{(\gamma)}(d-x)^{(\varepsilon)}}\right], \tag{50h}$$

$${}_{1+u}F_u\left({-s,\ (a_i+x+1)_p,\ (x-b_j+\beta_j)_q,\ (x+c_k-\gamma_k+1)_r,\ (x-d_l)_t \atop (a_i+x-\alpha_i+1)_p,\ (x-b_j)_q,\ (x+c_k+1)_r,\ (x-d_l+\delta_l)_t};\ z\right)=$$

$$=(-1)^s\left[z^x\frac{\prod_{i=1}^{p}(a_i+x)^{(\alpha_i)}\prod_{j=1}^{q}(b_j-x)^{(\beta_j)}}{\prod_{k=1}^{r}(c_k+x)^{(\gamma_k)}\prod_{l=1}^{t}(d_l-x)^{(\delta_l)}}\right]^{-1}\Delta_x^s\left[z^x\frac{\prod_{i=1}^{p}(a_i+x)^{(\alpha_i)}\prod_{j=1}^{q}(b_j-x)^{(\beta_j)}}{\prod_{k=1}^{r}(c_k+x)^{(\gamma_k)}\prod_{l=1}^{t}(d_l-x)^{(\delta_l)}}\right],\quad u=p+q+r+t. \tag{50i}$$

These equations are proved in the same way as Eqs. (49).

These examples show how to construct the CR representation of arbitrary ${}_pF_q$ functions with the aid of complicated generating functions $\Psi(x)$ made up of "elementary" $\Psi_i^0(x)$ and the results shown in Eq. (49). The following general rule exists for determining the form of the ${}_pF_q$ functions produced by a generating function of the form $\Psi(x)=\prod\Psi_i^0(x)$: each "elementary" factor $\Psi_i^0(x)$ "creates" the same "discrete" parameter a_i or b_j, or factor with the "continuous" argument z in the function ${}_pF_q\left({\cdots \atop \cdots};\ \ldots\right)$ as in the "free" state (except for the general parameter s "produced" by the operator Δ_x^s). From these examples it can be seen that the form of the CR representation for ${}_pF_q$ functions of a single type is not unique. The transformation from one form to another is completed with the aid of Eqs. (30).

The CR representation offers a number of operational possibilities for manipulation with finite generalized hypergeometric series. We now enumerate some of them and illustrate them with some examples that are useful in practice.

1. Difference "differentiation" of a ${}_pF_q$ function with respect to the "discrete" variables. If the "discrete" variable with respect to which the "differentiation" is being done is contained in one parameter, then the result is easy to obtain by applying Δ_x^s to ${}_pF_q\left({\cdots \atop \cdots};\ z\right)$ in the form

of Eq. (45) and using Eqs. (29a) and (30). For example, we have

$$\Delta_x^s\, {}_{p+1}F_q\left({(a_i)_p,\ c+x \atop (b_j)_q};\ y\right) = (-1)^{sp-sq}\, y^s \frac{\prod_{i=1}^{p}(-a_i)^{(s)}}{\prod_{j=1}^{q}(-b_j)^{(s)}}\, {}_{p+1}F_q\left({(a_i+s)_p,\ c+x+s \atop (b_j+s)_q};\ y\right); \tag{51a}$$

$$\Delta_x^s\, {}_{p+1}F_q\left({(a_i)_p,\ c-x \atop (b_j)_q};\ y\right) = (-1)^{s+sp-sq} y^s \prod_{i=1}^{p}(-a_i)^{(s)} \prod_{j=1}^{q}\frac{1}{(-b_j)^{(s)}}\, {}_pF_q\left({(a_i+s)_p,\ c-x \atop (b_j+s)_q};\ y\right). \tag{51b}$$

However, for more general dependences of several or all the parameters on the "discrete differentiation" parameter, it is appropriate to use the CR representation, which allows us to obtain the result almost automatically. Thus, for example applying Δ_x^p to the function ${}_3F_2\left({a+x+1,\ x-b+\beta,\ -s \atop a+x-\alpha+1,\ x-b};\ 1\right)$, written in the form of Eq. (50), and using Eqs. (26) and (50e), we easily find that [35]

$$\Delta_x^p\, {}_3F_2\left({a+x+1,\ x-b+\beta,\ -s \atop a+x-\alpha+1,\ x-b};\ 1\right) = \sum_{m,n} \frac{(-1)^{m+n}\, p!}{(m+n-p)!\,(p-m)!\,(p-n)!} \times$$

$$\times\, {}_3F_2\left({a+x-\alpha+1,\ x-b,\ -n \atop a+x+1,\ x-b+\beta};\ 1\right) {}_3F_2\left({a+x+1,\ x+\beta-b,\ -(s+m) \atop a+x-\alpha+1,\ x-b};\ 1\right). \tag{52}$$

We also note that Eq. (43) may be useful when difference "differentiating" a ${}_pF_q$ function in the CR representation.

2. Summation theorems, recurrence and other relations. Various relationships among ${}_pF_q$ functions play an important role in the theory and applications of hypergeometric functions. These allow us both to simplify computational schemes and to easily establish certain nontrivial facts [37, 38]. The CR representation of ${}_pF_q$ functions together with Eqs. (26), (42), and (43) offer practically unlimited possibilities of deriving various relations of this type. We now give some examples.

Setting z = 1 in Eq. (50i) and using Eq. (42) to calculate $\Delta_x^s[\ldots]$, we obtain the following:

$${}_{u+1}F_u\left({-s,\ (a_i+x+1)_p,\ (x-b_j+\beta_j)_q,\ (x+c_k-\gamma_k+1)_r,\ (x-d_l)_t \atop (a_i+x-\alpha_i+1)_p,\ (x-b_j)_q,\ (x+c_k+1)_r,\ (x-d_l+\delta_l)_t};\ 1\right) =$$

$$= (-1)^s \sum_{\sigma_1,\sigma_2,\sigma_3,\sigma_4} {}_4F_3\left({(\sigma_m-s)_4 \atop (-s)_3};\ 1\right) (-1)^{\sum_{m=1}^{4}\sigma_m} \prod_{m=1}^{4} \frac{s^{(\sigma_m)}}{\sigma_m!} \times$$

$$\times\, {}_{p+1}F_p\left({-\sigma_1,\ (a_i+x+1)_p \atop (a_i+x-\alpha_i+1)_p};\ 1\right) {}_{q+1}F_q\left({-\sigma_2,\ (x-b_j+\beta_j)_q \atop (x-b_j)_q};\ 1\right) \times$$

$$\times\, {}_{r+1}F_r\left({-\sigma_3,\ (x+c_k-\gamma_k+1)_r \atop (a+c_k+1)_r},\ 1\right) {}_{1+t}F_t\left({-\sigma_4,\ (x-d_l)_t \atop (x-d_l+\delta_l)_t};\ 1\right)$$

$$= (-1)^s \sum_{\{\lambda_m\}} {}_uF_{u-1}\left({(\lambda_m-s)_u \atop (-s)_{u-1}};\ 1\right) \prod_{m=1}^{u} {}_2F_1\left({-\lambda_m,\ s-\lambda_m \atop -\lambda_m};\ 1\right) \times$$

$$\times (-1)^{\sum_{m=1}^{u}\lambda_m} \prod_{i=1}^{p} {}_2F_1\left({a_i+x+1,\ -\lambda_i \atop a_i+x-\alpha_i+1};\ 1\right) \prod_{j=1}^{q} {}_2F_1\left({x-b_j+\beta_j,\ -\lambda_{j+p} \atop x-\beta_j};\ 1\right) \times$$

$$\times \prod_{k=1}^{r} {}_2F_1\left({x+c_k-\gamma_k+1,\ -\lambda_{k+p+r} \atop x+c_k+1};\ 1\right) \prod_{l=1}^{t} {}_2F_1\left({x-d_l,\ -\lambda_{l+p+q+r} \atop x-d_l+\delta_l};\ 1\right), \tag{53}$$

$$u = p+q+r+t,$$

which when read from right to left may be regarded as a summation theorem and when read from left to right, as a formula for reducing higher generalized hypergeometric functions to lower. In particular cases a number of useful relations automatically follow from Eq. (53), for example:

$$ {}_3F_2\left(\begin{matrix} a+1,\ \beta-b,\ -s \\ a-\alpha+1,\ -b \end{matrix};1\right)=\sum_{k_1,k_2}\frac{(-1)^{s+k_1+k_2}s!}{(k_1+k_2-s)!\,(s-k_1)!\,(s-k_2)!}\,{}_2F_1\left(\begin{matrix}\beta-b,\ -k_1 \\ -b\end{matrix};1\right){}_2F_1\left(\begin{matrix}a+1,\ -k_2 \\ a-\alpha+1\end{matrix};1\right); \tag{54a} $$

$$ {}_5F_4\left(\begin{matrix} a+1,\ \beta-b,\ c-\gamma+1,\ -d,\ -s \\ c+1,\ -d+\varepsilon,\ a-\alpha+1,\ -b \end{matrix};1\right)=\sum_{m,n}\frac{s!\,(-1)^{m+n+s}}{(m+n-s)!\,(s-m)!\,(s-n)!}\times $$
$$ \times\ {}_3F_2\left(\begin{matrix}a+1,\ -b+\beta,\ -m \\ a+1-\alpha,\ -b\end{matrix};1\right){}_3F_2\left(\begin{matrix}c-\gamma+1,\ -d,\ -n \\ c+1,\ -d+\varepsilon\end{matrix};1\right). \tag{54b} $$

In a similar way it is also possible to obtain this type of relation when the "continuous" argument z has arbitrary values. Thus, for example, setting z = uv in Eq. (50c) and using Eq. (26) to evaluate $\Delta^s_x[\ldots]$, we find

$$ {}_3F_1\left(\begin{matrix}-s,\ a+x+1,\ x-b \\ a+x-\alpha+1\end{matrix};uv\right)=\sum_{n,k}\frac{(-1)^{k+n+s}s!}{(n+k-s)!\,(s-k)!\,(s-n)!}\,{}_2F_0\left(\begin{matrix}x-b,\ -n \\ \ \end{matrix};v\right){}_2F_1\left(\begin{matrix}a+x+1,\ -k \\ a+x-\alpha+1\end{matrix};u\right). \tag{55} $$

A number of other possibilities for obtaining recurrence relations are offered by the properties of symbolic power functions given in Eq. (30) and the identity $\Delta^{p+q}_x\equiv\Delta^p_x\Delta^q_x$. Thus, transforming $(a+x)^{(\alpha)}(b-x)^{(\beta)}$ in Eq. (50d) to the form $[(a+x)^{(k)}(b-x)^{(l)}][(a-k+x)^{(\alpha-k)}(b-l-x)^{(\beta-l)}]$ and using the above procedure, we obtain

$$ {}_3F_2\left(\begin{matrix}a+x+1,\ x-b+\beta,\ -s \\ a+x-\alpha+1,\ x-b\end{matrix};1\right)=\sum_{m,n}\frac{(-1)^{m+n+s}s!}{(m+n-s)!\,(s-n)!\,(s-m)!}\times $$
$$ \times\ {}_3F_2\left(\begin{matrix}a+x+1,\ x-b+l,\ -m \\ x-b,\ a+x-k+1\end{matrix};1\right){}_3F_2\left(\begin{matrix}a-k+x+1,\ x-b+\beta,\ -n \\ x-b+l,\ a+x-\alpha+1\end{matrix};1\right), \tag{56} $$
$$ 0\leqslant k\leqslant\alpha,\quad 0\leqslant l\leqslant\beta. $$

On the other hand, setting s = p + q in Eq. (50d), we find, using $\Delta^{p+q}_x=\Delta^p_x\Delta^q_x$ and Eqs. (26) and (52), that

$$ {}_3F_2\left(\begin{matrix}a+x+1,\ x-b+\beta,\ -(p+q) \\ x-b,\ a+x-\alpha+1\end{matrix};1\right)= $$
$$ =\sum_{\substack{k,n\\m,l}}\frac{(-1)^{k+l+m+p}n!\,p!}{(n+k-p)!\,(p-n)!\,(p-k)!\,(m+l-n)!\,(n-l)!\,(n-m)!}\times $$
$$ \times\ {}_3F_2\left(\begin{matrix}a+x+1,\ x+\beta-b,\ -(q+l) \\ a+x-\alpha+1,\ x-b\end{matrix};1\right)\times $$
$$ \times\ {}_3F_2\left(\begin{matrix}a+x-\alpha+1,\ x-b,\ -m \\ a+x+1,\ x-b+\beta\end{matrix};1\right){}_3F_2\left(\begin{matrix}a+x+1,\ a+\beta-b,\ -k \\ a+x-\alpha+1,\ x-b\end{matrix};1\right) $$
$$ =\sum_{m,k,l}p^{(k)}(-1)^{p+k+l+m}\,{}_3F_2\left(\begin{matrix}a+x+1,\ x-b+\beta,\ -k \\ a+x-\alpha+1,\ x-b\end{matrix};1\right)\times $$
$$ \times\ {}_3F_2\left(\begin{matrix}a+x-\alpha+1,\ x-\beta,\ -m \\ a+x+1,\ x-b+\beta\end{matrix};1\right){}_3F_2\left(\begin{matrix}a+x+1,\ x-b+\beta,\ -(q+l) \\ a+x-\alpha+1,\ x-b\end{matrix};1\right)\times $$
$$ \times\sum_n\frac{n!}{(n+k-p)!\,(p-n)!\,(m+l-n)!\,(n-l)!\,(n-m)!}. \tag{57} $$

Equations (56) and (57) have an interesting interpretation in the theory of angular momentum (see Section 3).

3. Expansion of arbitrary functions $\Psi(x)$ in power series. With the CR representation of ${}_pF_q$ functions it is easy to write factorial series [of the type of Eq. (31)] expansions of generating functions $\Psi(x)$ in powers $x^{(n)}$.

Thus, for example, we have

$$\prod_{i=1}^{p}(a_i+x)^{(\alpha_i)}\prod_{j=1}^{q}(b_j-x)^{(\beta_j)}=$$
$$=\sum_{l=0}^{\infty}\frac{(-1)^l}{l!}x^{(l)}\prod_{i=1}^{p}a_i^{(\alpha_i)}\prod_{j=1}^{q}b_j^{(\beta_j)}\,{}_{1+q+p}F_{p+q}\left({-l,\ (a_i+1)_p,\ (\beta_j-b_j)_q \atop (a_i-\alpha_i+1)_p\ (-b_j)_q};\ 1\right), \tag{58a}$$

$$b^{(x)}=\sum_{k=0}^{\infty}\frac{x^{(k)}}{k!}(-1)^k\,{}_2F_0\left(-b,\ -k;\ -1\right), \tag{58b}$$

$$[(a+x)^{(\alpha)}]^{-1}=\sum_{k=0}^{\infty}\frac{x^{(k)}}{k!}\frac{(-1)^k}{a^{(\alpha)}}\,{}_2F_1\left({a+1-\alpha,\ -s \atop a+1};\ 1\right), \tag{58c}$$

$$(a+x)^{(l)}(b-x)^{(r)}=\sum_{m=0}^{l+r}\frac{x^{(m)}}{m!}(-1)^m a^{(l)}b^{(r)}\,{}_3F_2\left({a+1,\ r-b,\ -m \atop a-l+1,\ -b};\ 1\right) \tag{58d}$$

(where l and r are natural numbers), etc. These series are discrete analogs of the expansions of generating functions for special functions in continuous analysis [37] and yield various relations upon manipulating with $\Psi(x)$. For example, applying the operator Δ_x^s to Eq. (58d), we obtain

$$\Delta_x^s[(a+x)^{(l)}(b-x)^{(r)}]=\sum_{k=0}^{r+l-s}\frac{x^{(k)}}{k!}(-1)^{k+s}a^{(l)}b^{(r)}\,{}_3F_2\left({a+1,\ r-b,\ -(s+k) \atop a-l+1,\ -b};\ 1\right) \tag{59}$$

or, recalling Eq. (50d),

$$(a+x)^{(l)}(b-x)^{(r)}\,{}_3F_2\left({a+x+1,\ x-b+r,\ -s \atop a+x-l+1,\ x-b};\ 1\right)=\sum_{k=0}^{l+r-s}\frac{x^{(k)}}{k!}(-1)^k a^{(l)}b^{(r)}\,{}_3F_2\left({a+1,\ r-b,\ -(s+k) \atop -b,\ a-l+1};\ 1\right). \tag{60}$$

When $s = l + r$ it follows from Eq. (59) that

$$\Delta_x^{l+r}[(a+x)^{(l)}(b-x)^{(r)}]=(-1)^{l+r}a^{(l)}b^{(r)}\,{}_3F_2\left({a+1,\ r-b,\ -(l+r) \atop -b,\ a-l+1};\ 1\right). \tag{61}$$

Many other results similar to Eqs. (53)-(57) may be obtained in the same way as in continuous analysis, by multiplying series of the type in Eq. (58) for various generating functions $\Psi(x)$, expanding both sides of the resulting expression in factorial series, and equating the coefficients for equal powers $x^{(n)}$ (or applying the operator $\Delta_x^s f(x)|_{x=0}$). We shall also show that with Stirling numbers of the first kind [see Eq. (68)] it is possible to go from factorial series for the generating functions $\Psi(x)$ to ordinary power series, thereby "transforming" the $\Psi(x)$ into ordinary generating functions (for weighted sums of ${}_pF_q$ functions).

4. Orthogonal polynomials. There is great interest in discretizing the ${}_pF_q$ functions and in writing them in the CR representation for the theory and applications of orthogonal polynomials. The classical systems of orthogonal polynomials in continuous analysis are defined by the generalized Rodriguez formula (48) and expressed in terms of Gaussian series of a particular type [37]. These polynomials have an important applied role since they are the solutions of several important classes of problems in mathematical physics. In connection with the rising number of physical problems in which a difference formulation develops uniquely (see Sections 1, 3, and 4), it is desirable to obtain the discrete analogs of the orthogonal polynomials of continuous analysis (and then in a form which allows at least a partial transfer of the methods of continuous analysis to discrete analysis). At present several systems $\{p_n(x)\}$

of orthogonal polynomials of a discrete variable (discrete orthogonal polynomials) [37] are known which are defined with the aid of a difference analog of Eq. (48),

$$p_n(x) = \frac{\Delta_x^n [j(x-n) X^{[n]}(x)]}{K_n j(x)}, \tag{62}$$

where j(x) is a discrete weighting function which is used to define the discrete variant of the scalar product $(p_n(x), p_m(x)) = \sum_x j(x) p_n(x) p_m(x)$, and $X^{[n]}(x) = X(x) X(x-1) \ldots X(x-n+1)$, where X(x) is a polynomial whose coefficients are independent of n (it is easy to show that $X^{[n]}(x)$ can be expressed in the form $X^{[n]}(x) = \prod_{i=1}^{p} (x-\alpha_i)^{(l_i)} \prod_{j=1}^{q} (\beta_j - x)^{(r_j)}$). Of the known orthogonal polynomials of a discrete variable the Hahn polynomials [37] have the greatest practical interest (see Section 3). These have the form

$$p_n(x;\, \beta,\, \gamma,\, \delta) = \frac{x!\,(\delta+x-1)^{(x)}}{(x+\beta-1)^{(x)}(\gamma+x-1)^{(x)}} \Delta_x^n \left[\frac{(\beta+x-1)^{(x)}(\gamma+x-1)^{(x)}}{(x-n)!\,(\delta-x-n-1)^{(x-n)}}\right], \tag{63}$$

which is the difference analog of the usual Jacobi polynomials and can be expressed in terms of the ${}_3F_2$ function as

$$p_n(x;\, \beta,\, \gamma,\, \delta) = \frac{(\beta-1+n)^{(n)}(\gamma-1+n)^{(n)}}{n!}\, {}_3F_2\left(\begin{matrix}-n,\ -x,\ \beta+\gamma-\delta+n \\ \beta,\ \gamma\end{matrix};\, 1\right). \tag{64}$$

In the particular case $\beta = 1, \gamma = 1-N, \delta = 1-N$ the Hahn polynomials degenerate to the discrete Chebyshev polynomials $t_n(x, N)$ [37], which occupy an important place in the theory of discrete approximations to functions and in smoothing out of measurement results by the method of least squares [27, 29, 37].

We note that Eq. (64), which was first established by Weber and Erdelyi [37], follows automatically from Eq. (63) and the CR representation of the ${}_pF_q$ functions. Thus it is possible to hope that the CR representation of the ${}_pF_q$ functions will not only make it simpler to prove the known results but will aid in obtaining new results, particularly in expanding the list of physically important orthogonal polynomials of a discrete variable.

A useful instrument for solving this problem (finding discrete orthogonal polynomials in the class of generalized hypergeometric functions with the aid of the CR representation of the latter) is Abel's identity (27). This expansion of the list of orthogonal polynomials is especially important for the theory of orthogonal polynomials in m ($m \geq 2$) variables which is still not adequately developed even in the framework of continuous analysis [37]. Meanwhile, the practical significance of these orthogonal polynomials is constantly growing due to the increased attention now being devoted to research on complex, multiparticle systems. Then the operation of discretizing the objects of continuous analysis (as opposed to the ${}_pF_q$ functions) may lead to new variables in the theory. Thus, for example, making the orthogonal Appell polynomials in two variables $F_{m,n}(\gamma, \gamma'; x, y)$ (defined by

$$F_{m,n}(\gamma,\, \gamma';\, x,\, y) = \frac{x^{1-\gamma} y^{1-\gamma'}}{(\gamma-1-m)^{(m)}(\gamma'-1+n)^{(n)}} \frac{\partial^{m+n}}{\partial x^m \partial y^n} [x^{\gamma+m-1} y^{\gamma'+n-1} (1-x-y)^{n+m}] \tag{65}$$

and expressed in terms of the generalized Appell hypergeometric function $F_2(-m-n, \gamma+m, \gamma'+n, \gamma, \gamma'; x, y)$, [37]) discrete by applying Eq. (33), we find that the discrete analog,

$$\tilde{F}_{m,n}(\gamma,\, \gamma';\, x,\, y)_{x,y} = \frac{x^{(1-\gamma)} y^{(1-\gamma')}}{(\gamma-1-m)^{(m)}(\gamma'-1+n)^{(n)}} \Delta_x^m \Delta_y^n [x^{(\gamma+m-1)} y^{(\gamma'+n-1)} (1-x-y)^{(m+n)}] =$$

$$= \frac{x^{(1-\gamma)} y^{(1-\gamma')} (m+n)!}{(\gamma-1+m)^{(m)}(\gamma'-1+n)^{(n)}} (-1)^{m+n} \left| \sum_{\alpha,\beta} \frac{x^{(\alpha)} y^{(\alpha)} (m+\alpha)^{(m+\gamma-1)} (n+\beta)^{(n+\gamma'-1)}}{\alpha!\,\beta!\,(\alpha+\beta+m+n-2)!} (2m+2n+\gamma+\gamma'-4)^{(\alpha+\beta+m+n-2)}, \right. \tag{66}$$

is not expressed in terms of the Appell hypergeometric functions in two variables $F_1 - F_4$, but, clearly, requires the use of more general hypergeometric functions (Horn–Birkeland, Horn, and Laurincella series [37]).

Without dwelling in detail on all the possibilities for expanding the theory of orthogonal polynomials (both with discrete and with continuous arguments) which are made possible by the CR representation of the ${}_pF_q$ functions (not their generalizations), we shall point out the following. Equations such as (51)-(61) established with the CR representation make it easy to expand various orthogonal polynomials (and their products) in terms of one another as well as of other ${}_pF_q$ functions (not necessarily reducible to orthogonal polynomials) in terms of systems of orthogonal polynomials. Since the CR representation of a ${}_pF_q$ function expresses the associated orthogonal polynomials all at once in the form of Eq. (62), it is possible, by analogy with continuous analysis and using Eq. (26) (and other equations in Section 2.1), to obtain the difference equations associated with each system of discrete orthogonal polynomials [37]. These results in combination with Abel's identity may be useful in analyzing and solving special difference equations generated by physical problems. Besides the already noted analogies and interrelations between discrete and continuous mathematics, it is possible that this will signify a new stage in the development of mathematical physics.

As a conclusion to this section we note that the above results and examples already indicate the rich prospects for using generalized hypergeometric functions in discrete analysis. However, these results may be expanded and generalized. In particular, the CR representations obtained above for finite hypergeometric series can evidently be extended in a number of cases even to arbitrary ${}_pF_q$ functions using the operators Δ_x^{α} (where α is an arbitrary real number) defined by Eq. (39d). Operators of the form ${}_pF_q\left(\begin{matrix}(a_i)_p\\(b_j)_q\end{matrix}; \Delta_y\right)$ offer interesting operational prospects in the field of discrete analysis, as is shown by the two formulas

$$ {}_pF_q\left(\begin{matrix}(a_i)_p\\(b_j)_q\end{matrix}; \Delta_y\right)(c+y)^{(\alpha)} = (c+y)^{(\alpha)}\, {}_{p+1}F_{q+1}\left(\begin{matrix}(a_i)_p,\ -\alpha\\(b_j)_q,\ c+y-\alpha+1\end{matrix}; -1\right), \tag{67a} $$

$$ {}_pF_q\left(\begin{matrix}(a_i)_p\\(b_j)\end{matrix}; \Delta_x\right) {}_{r+1}F_r\left(\begin{matrix}(c_j+x)_r,\ -s\\(c_j+x-a_j)_r\end{matrix}; 1\right) = \sum_{n,m,k} \frac{(-1)^{np-nq+n}}{(n-k)!\,(n-m)!\,(k+m-n)!} \cdot \frac{\prod_{i=1}^{p}(-a_i)^{(n)}}{\prod_{j=1}^{q}(-b_j)^{(n)}} \times $$

$$ \times\, {}_{r+1}F_r\left(\begin{matrix}(c_j+x)_r,\ -(m+s)\\(c_j+x-a_j)_r\end{matrix}; 1\right) {}_{r+1}F_r\left(\begin{matrix}(c_j+x-a_j)_r,\ -k\\(c_j+x)_r\end{matrix}; 1\right), \tag{67b} $$

whose validity may be verified by direct calculation [keeping definition (44) in mind]. All this leads us to conclude that generalized hypergeometric functions and the finite difference formalism may be used as a basis for creating a powerful, structured calculus for solving many problems in contemporary physics.

2.3. The Stirling Numbers

We shall devote the last portion of this section to the Stirling numbers, which are of interest from the standpoint of the interrelation between discrete and continuous analysis and in applications [29]. Thus, it is desirable to have formulas for evaluating them numerically in addition to the customary implicit definitions in terms of recurrence relations (difference equations) and statements of their simplest properties. These equations and some other practically useful relations which contain Stirling numbers will be derived below with the aid of the properties of the Δ_x operator and its functions. In this way several of the theorems of Section 2.1 will be illustrated.

The Stirling numbers of the first, s(n, r), and second, S(n; r), kinds are defined by the relations

$$x^{(n)} \equiv \sum_{r=0}^{n} s(n, r) x^{r}, \tag{68a}$$

$$x^{n} \equiv \sum_{r=0}^{n} S(n; r) x^{(r)} \tag{68b}$$

(where n is a natural number), and are the expansions of symbolic powers in terms of ordinary powers, and vice versa, respectively. From the definitions (68) and (28) we easily obtain numerical values of s(n, r) and S(n; r) for a number of special values of n and r,

$$s(n, 0) = S(n; 0) = \delta_{n,0}, \tag{69a}$$

$$s(1, r) = S(1; r) = \delta_{1,r}, \tag{69b}$$

$$s(n, n) = S(n; n) = 1, \tag{69c}$$

$$s(n, n+k) = S(n; n+k) = 0 \quad \text{for} \quad k > 0, \tag{69d}$$

where $\delta_{a,b}$ is the Kronecker delta, and obtain the orthogonality relations

$$\sum_{l} S(n; l) s(l, m) = \delta_{n,m}, \tag{70a}$$

$$\sum_{l} s(n, l) S(l; m) = \delta_{n,m} \tag{70b}$$

for the two systems $\{S(n; l)\}$, $\{s(n, l)\}$ of numbers (thereby allowing us to treat them as two biorthogonal bases). In the general case of arbitrary n and r the numerical values of s(n, r) and S(n; r) may be found using the relationships [28] between them,

$$s(n, p) = \frac{(n-1)^{(n-p)}}{(n-p)!} B_{n-p}^{(n)}, \tag{71a}$$

$$S(n, p) = \frac{n^{(p)}}{p!} B_{n-p}^{(-p)}, \tag{71b}$$

and the Bernoulli numbers of the n-th (and −p-th) orders, respectively (n, p = 0, 1, ...). The latter are defined in terms of the generating functions as follows [28, 37]:

$$\left[\frac{z}{e^{z}-1}\right]^{\mu} = \sum_{\nu=0}^{\infty} \frac{z^{\nu}}{\nu!} B_{\nu}^{(\mu)}, \tag{72}$$

where μ is an arbitrary whole number. From the standpoint of obtaining definitive results in the form of an equation, however, it is best to turn to computational schemes based on the definitions of Eq. (69) or on certain "defining" relations which follow from them. The standard defining relations are [29]

$$s(n+1, r+1) = s(n, r) - ns(n, r+1), \tag{73a}$$

$$S(n+1; r+1) = S(n; r) + (r+1) S(n; r+1), \tag{73b}$$

which are derived from Eq. (68) with the aid of an elementary identity for

$$x^{(r)} x \equiv x^{(r)} x^{(1)} = x^{(r+1)} + r x^{(r)}. \tag{74}$$

We now introduce two basic computational schemes, combinations of which may be used to obtain many others.

1. A "direct" way of calculating s(n, r) and S(n; r). Equation (68a) is a Maclaurin series for $x^{(n)}$ while Eq. (68b) is its difference analog. Thus, we have

$$S(n;\ r) \equiv \frac{1}{r!} \Delta_x^r x^n \big|_{x=0} \tag{75}$$

and

$$s(n,\ r) \equiv \frac{1}{r!} D_x^r x^{(n)} \big|_{x=0} \quad (D_x \equiv d/dx) \tag{76}$$

as computational schemes (constructional definitions) for S(n; r) and s(n, r) respectively. Then Eq. (75) yields the result very rapidly. Using (24), we find

$$S(n;\ r) \equiv \frac{1}{r!} \Delta_x^r x^n \big|_{x=0} = \sum_{l=0}^{r} (-1)^{l+r} \frac{l^n}{l!\,(r-l)!} = \sum_{l=0}^{r} (-1)^l \frac{(r-l)^n}{l!\,(r-l)!}. \tag{77}$$

The final result for s(n, r) cannot be obtained so directly with Eq. (76) and may be derived in various specific ways. One of these uses the formal relation [27-30]

$$\Delta_x \equiv V_x - 1 \equiv \exp D_x - 1, \tag{78}$$

which is valid for application of Δ_x to differentiable functions, and yields [including Eq. (39c)] the following chain of formal equations:

$$s(n,\ r) \equiv \frac{1}{r!} D_x^r x^{(n)} \big|_{x=0} = \frac{1}{r!} [\ln(1+\Delta_x)]^r x^{(n)} \big|_{x=0} = \frac{1}{r!} \left[\sum_{k=1}^{\infty} (-1)^{k-1} \frac{\Delta_x^k}{k} \right]^r x^{(n)} \big|_{x=0}$$

$$= \frac{1}{r!} \left[\sum_{k=0}^{\infty} (-1)^k \frac{\Delta_x^{k+1}}{k+1} \right]^r x^{(n)} \big|_{x=0}. \tag{79}$$

We now use the fact that $\Delta_x^k x^{(n)} = 0$ when k > n. Then we find (successively applying the series $\sum_k \frac{(-1)^k}{k+1} \Delta_x^{k+1}$ to $x^{(n)}$ r times) an equation for s(n, r):

$$s(n,\ r) = (-1)^{n-r} \frac{n!}{r!} \sum_{k_1=0}^{n-r} \sum_{k_2=0}^{n-r-k_1} \cdots \sum_{k_{r-1}=0}^{n-r-k_1-\ldots-k_{r-2}} \frac{1}{(k_1+1)(k_2+1)\ldots(k_{r-1}+1)(n-r-k_1-\ldots-k_{r-1}+1)}. \tag{80}$$

On the other hand, leaving only the first n − r + 1 terms in $\sum_{k=1}^{\infty} (-1)^{k-1} \frac{\Delta_x^k}{k}$ and using the formula for a polynomial expansion, we ar. ve at the following result:

$$s(n,\ r) = (-1)^{n-r} \sum_{\substack{\{\alpha_i \geqslant 0\} \\ \alpha_1+\alpha_2+\ldots+\alpha_{n-r+1}=r \\ \alpha_1+2\alpha_2+\ldots+(n-r+1)\alpha_{n-r+1}=n}} \frac{n!}{\prod_{i=1}^{n-r+1} \alpha_i!\, i^{\alpha_i}}. \tag{81}$$

(It is interesting to note that under the summation sign in Eq. (81) there are a number of elements in a class with the cyclic structure $(\alpha) = (1^{\alpha_1},\ 2^{\alpha_2},\ \ldots,\ (n-r+1)^{\alpha_{n-r+1}}, (n-r+2)^0, \ldots)$ of the S_n group of permutations of n objects [39].)

The same result [Eqs. (80) and (81)] may be obtained by evaluating $D_x x^{(n)}$ in Eq. (76) according to the definition

$$D_x x^{(n)} \equiv \lim_{\Delta x \to 0} \frac{1}{\Delta x} [(x+\Delta x)^{(n)} - x^{(n)}] = \sum_{\beta=1}^{n} \frac{n^{(\beta)}}{\beta!} (-1)^{(\beta-1)} x^{(n-\beta)} = \sum_{\beta=1}^{n} \frac{n^{(\beta)}}{\beta} (-1)^{\beta-1} x^{(n-\beta)} \equiv \sum_{\beta=1}^{\infty} \frac{(-1)^{\beta-1}}{\beta} \Delta_x^{\beta} x^{(n)}. \tag{82}$$

[Equations (32a) and (30a) were used in deriving Eq. (82).]

A comparison of Eq. (82) with Eq. (79) proves the validity of the statement made above: Eq. (82) may be regarded as a direct basis for representing $D_x = \ln(1+\Delta_x)$ by the formal power series (39c) in the ring of polynomials in $x^{(n)}$. We can obtain yet another concrete realization of the scheme of Eq. (76) by using the representation $x^{(n)} \equiv x(x-1)\ldots(x-n+1) = \prod_{s=0}^{n-1}(x-s)$. We finally find

$$s(n, r) \equiv \frac{1}{r!} D_x^r x^{(n)} \Big|_{x=0} = \frac{1}{r!} D_x^r \prod_{s=0}^{n-1} (x-s) \Big|_{x=0} = \sum_{\substack{\{\alpha_i \geqslant 0\} \\ \alpha_0+\alpha_1+\ldots+\alpha_{n-1}=r}} \prod_{s=0}^{n-1} \left[\frac{1}{\alpha_s!} D_x^{\alpha_s}(x-s)\right]\Big|_{x=0} =$$

$$= (-1)^{n-r}(n-1)! \sum_{\substack{\{0 \leqslant \alpha_i \leqslant 1\} \\ \alpha_1+\alpha_2+\ldots+\alpha_{n-1}=r-1}} \left[\prod_{l=1}^{n-1} l^{\alpha_l}\right]^{-1}, \quad n \geqslant r > 0. \tag{83}$$

2. Evaluating s(n, r) and S(n; r) by solving the defining equations (73). Setting $n = r + k$ ($k \geq 0$) in Eq. (73), we obtain

$$s(k+r+1, r+1) = s(k+r, r) - (k+r)\, s(k+r, r+1), \tag{84a}$$

$$S(k+r+1; r+1) = S(k+r; r) + (r+1)\, S(k+r; r+1). \tag{84b}$$

Introducing the notations $s(k+r, r) \equiv \omega(k, r)$ and $S(k+r; r) \equiv \Omega(k, r)$, we transform Eq. (84) to the form

$$\Delta_r \omega(k, r) = -(k+r)\, \omega(k-1, r+1); \tag{85a}$$

$$\Delta_r \Omega(k, r) = (r+1)\, \Omega(k-1, r+1). \tag{85b}$$

Then, using Eq. (25) and condition (69), we find

$$\omega(k, r) = -\sum_{l=0}^{r-1} (l+k)\, \omega(k-1, l+1) + \delta_{k,0} = -\sum_{m=1}^{r} (m+k-1)\, \omega(k-1, m) + \delta_{k,0}, \tag{86a}$$

$$\Omega(k, r) = \sum_{l=0}^{r-1} (l+1)\, \Omega(k-1, l+1) + \delta_{k,0} = \sum_{l=1}^{r} l\, \Omega(k-1, l) + \delta_{k,0}. \tag{86b}$$

In the original notation Eqs. (86) take the form

$$s(k+r, r) = -\sum_{l=1}^{r} (l+k-1)\, s(k+l-1, l) + \delta_{k,0}, \tag{87a}$$

$$S(k+r; r) = \sum_{l=1}^{r} l\, S(k+l-1; l) + \delta_{k,0}. \tag{87b}$$

Equations (87) are linear computational schemes which express s(n, r) and S(n; r), respectively, in terms of the values of similar quantities having smaller values of n and r. By successively applying these reducing schemes several times and having explicit expressions for the numbers s(k, l) and S(k; l) entering in the final summation, it is possible to obtain expressions for s(n, r) and S(n; r) in the form of iterated sums of completely defined expressions (functions). Thus, using this reduction scheme together with Eq. (69), we obtain the following expressions for s(n, r) and S(n; r):

$$s(n, r) = (-1)^{n-r}(1-\delta_{n,r}) \sum_{l_1=1}^{r} \sum_{l_2=1}^{l_1} \cdots \sum_{l_{n-r}=1}^{l_{n-r-1}} \prod_{i=1}^{n-r} (l_i + n - r - i) + \delta_{n,r}, \tag{88a}$$

$$S(n; r) = (1-\delta_{n,r}) \sum_{l_1=1}^{r} \sum_{l_2=1}^{l_1} \cdots \sum_{l_{n-r}=1}^{l_{n-r-1}} \prod_{i=1}^{n-r} l_i + \delta_{n,r}. \tag{88b}$$

Recalling Eqs. (29c) and (74), it is possible (successively getting rid of the inner sums) to obtain numerical values for s(n, r) and S(n; r) in the form of factorial series.

The general equations (77), (80), (81), and (88) found above may be used both to tabulate directly the numerical values of s(n, r) and S(n; r) and to derive special formulas.

As an illustration of this discussion and for reference we now write a number of such formulas. For s(n, k), we have (n ≥ k):

$$s(n, 1) = (-1)^{n-1}(n-1)!, \tag{89a}$$

$$s(n, 2) = (-1)^{n-2}(n-1)! \sum_{k=0}^{n-2} \frac{1}{k+1}, \tag{89b}$$

$$s(n, 3) = (-1)^{n-3} n! \sum_{k=0}^{n-3} \frac{1}{(k+1)(n-k-1)} \sum_{l=0}^{n-3-k} \frac{1}{l+1}, \tag{89c}$$

$$s(n, 4) = \frac{n!}{2 \cdot 3!} (-1)^{n-4} \sum_{\alpha=1}^{n-3} \sum_{\beta=1}^{n-2-\alpha} \sum_{\gamma=1}^{n-1-\alpha-\beta} \frac{1}{\alpha\beta\gamma(n-\alpha-\beta)}, \tag{89d}$$

$$s(n, n-1) = -\frac{n^{(2)}}{2}, \tag{89e}$$

$$s(n, n-2) = \frac{(n+1)^{(4)}}{2^3} - \frac{n^{(3)}}{3!}, \tag{89f}$$

$$s(n, n-3) = -\frac{1}{4!}\left[\frac{(n+2)^{(6)}}{2} - 2(n+1)^{(5)} + n^{(4)}\right]. \tag{89g}$$

Some formulas for S(n; k) (n ≥ k):

$$S(n; 1) = 1, \tag{90a}$$

$$S(n; 2) = 2^{n-1} - 1, \tag{90b}$$

$$S(n; 3) = \frac{1}{2}[3^{n-1} - 2^n + 1], \tag{90c}$$

$$S(n; 4) = \frac{1}{3!}[4^{n-1} - 3^n + 3 \cdot 2^{n-1} - 1], \tag{90d}$$

$$S(n;\ n-1)=n^{(2)}/2, \tag{90e}$$

$$S(n;\ n-2)=-\frac{n^{(3)}}{3}+\frac{(n+1)^{(4)}}{2^3}, \tag{90f}$$

$$S(n;\ n-3)=\frac{1}{4!}\left[n^{(4)}+2n^{(5)}+\frac{n^{(6)}}{2}\right]; \tag{90g}$$

$$S(n;\ n-4)=\frac{n^{(5)}}{5!}+\frac{5n^{(6)}}{4!3!}+\frac{n^{(7)}}{2\cdot 4!}+\frac{n^{(8)}}{2^4 4!}. \tag{90h}$$

These formulas for s(n, k) and S(n; k) may be a source of numerous combinatorial identities and practically useful expressions which proceed from comparison of various expressions for equivalent quantities and from the elementary properties of s(n, k), S(n; k), and the symbolic powers. Without dwelling on this in detail, we present just a few examples.

From Eqs. (77) and (69d) it follows that

$$\sum_{k=1}^{n}(-1)^k\frac{(n+1-k)^{n-1}}{k!\,(n-k)!}=-\frac{(n+1)^{n-1}}{n!}. \tag{91}$$

With Eq. (36) we obtain

$$\sum_{\substack{m\geqslant l\geqslant 0\\ n\geqslant k\geqslant 0\\ k+l\geqslant r}} S(k+l;\ r)\,s(n,\ k)\,s(m,\ l)=\frac{m!\,n!}{(r-m)!\,(r-n)!\,(m+n-r)!}, \tag{92}$$

from which, with Eq. (70), we find

$$\sum_{l=r-p}^{m} S(l+p;\ r)\,s(m,\ l)=\sum_{n}\frac{m!\,n!\,S(p;\ n)}{(r-m)!\,(r-n)!\,(m+n-r)!}. \tag{93}$$

From the binomial expansions of $(x+y)^n$ in ordinary powers and of $(x+y)^{(n)}$ in symbolic powers, together with Eq. (68), we have

$$\sum_{\beta=0}^{n-l}\frac{n^{(\beta)}}{\beta!}S(n-\beta;\ l)\,S(\beta;\ \alpha)=\frac{(l+\alpha)^{(l)}}{l!}S(n;\ l+\alpha), \tag{94a}$$

$$\sum_{\beta=0}^{r-k}\frac{r^{(\beta)}}{\beta!}s(r-\beta,\ k)\,s(\beta,\ m)=\frac{(k+m)!}{k!\,m!}s(n,\ k+m). \tag{94b}$$

As a conclusion to this section, we note that the results obtained here may be used to do specific calculations when going from discrete to continuous analysis, and vice versa. In this regard, to supplement the remarks of Sections 2.1 and 2.2, we note in particular that with the relations [28]

$$B_n\equiv\sum_{m=0}^{n}(-1)^m\frac{m!}{m+1}S(n;\ m) \tag{95}$$

Eqs. (77), (88), and (90) may be used to evaluate ordinary (first kind) Bernoulli numbers B_n that are used in the Euler–Maclaurin formula [27],

$$[P^{-1}_{(r)}-P^{-1}_{(m)}]\{u(x)\}\equiv\sum_{x=m}^{r-1}u(x)=\int_m^r u(x)\,dx+\sum_{\nu=1}^{\infty}\frac{B_\nu}{\nu!}D_x^{\nu-1}u(x)|_m^r, \tag{96}$$

which replaces the discrete operation $P^{-1}_{(x)}$ by the continuous D_x and $\int dx$.

With this we conclude our analysis of the special problems of discrete analysis and proceed to discuss the prospects for using the results in developing the "symmetry" formalism.

3. Hypergeometric Functions, Orthogonal Polynomials, and the Theory of Angular Momentum

In Section 1 we pointed out the possibility of developing the theory of coherent cooperative phenomena using a "symmetry" formalism based on representations of various groups of physical symmetries, primarily the SU_n group. The basic elements of this formalism are (1) the realization of irreducible representations of groups (specifying the bases and the matrix elements of the generators and other operators of the irreducible representations) in physical terms, i.e., elementary "symmetric" models of physical objects, and (2) the theory of Clebsch–Gordan coefficients and their combinations which arise naturally in the analysis of complex systems [40]. At present the first aspect has been developed fairly adequately for most groups with physical applications; however, the theory of Clebsch–Gordan coefficients for physical symmetry groups is still far from complete (see, for example, [33, 34, 40] and the references cited there). Only the theory (more precisely, its classical variant) of the Clebsch–Gordan coefficients of the SU_2 group (and closely related groups), that is, the theory of angula momentum [41, 42], is relatively complete. As for the other groups there are a number of problems. Thus, difficulties arise with the Clebsch–Gordan coefficients of unitary irreducible representations of noncompact groups because of the infinite dimensionality of such representations. For higher compact groups, SU_n $(n>2)$, SO_m $(m>3)$, Sp_{2l} $(l \geq 2)$, the theory of the Clebsch–Gordan coefficients is in many respects similar to the higher orders of the theory of angular momentum (generalized Clebsch–Gordan coefficients, the 3nj-symbols). In particular, this has made it possible to carry over from the latter the elements of the Wigner–Racah algebraic techniques for constructing arbitrary quantities in the theory of Clebsch–Gordan coefficients of higher groups from a few basic quantities. There are, however, substantial differences between the theory of angular momentum and the theory of the Clebsch–Gordan coefficients of higher groups. This is primarily due to the fact that for higher groups with a direct physical significance, the canonical (orthonormal) bases of an irreducible representation are not (as opposed to the case of the SU_2 group) defined uniquely and depend on the type of physical problem. (A formal expression of this ambiguity is the differences in the chain of subgroups of a higher group which defines the basis [34, 40].) The transformations among Clebsch–Gordan coefficients calculated in various bases is far from simple, a fact which makes the analytic apparatus for the theory of the Clebsch–Gordan coefficients of higher groups much more complicated.

This situation, combined with the considerable extension of the list of higher compact groups with diverse physical applications, poses the problem of creating a unified theory of Clebsch–Gordan coefficients of compact groups using some (bounded) general set basis quantities and operations, and of creating with them the Clebsch–Gordan coefficients (and their combinations) for all (or, at least, for large classes of) these groups. Using the factorizability of the Clebsch–Gordan coefficients of higher groups (the separation of G_i scalar factors, where G_i are elements of a chain of subgroups defining the canonical bases), it has been proposed that the basis elements of the theory of angular momentum (the Clebsch–Gordan coefficients of SU_2 [34, 40]) be taken as the basis quantities of such a unified theory.

Certain successes have been achieved in this direction; in particular, it has been possible to translate the theory of the Clebsch–Gordan coefficients of the SU_n $(n \geq 3)$ group into the theory of angular momentum. However, in doing this a number of problems have arisen, in particular, the problem of nonstandard summation of the Clebsch–Gordan coefficients of SU_2 and other quantities from the theory of angular momentum when the Clebsch–Gordan coefficients of higher groups are being constructed from then. These difficulties have been partially overcome by using the Regge symmetry [43] of the Clebsch–Gordan and Racah coefficients of the SU_2 group. On the whole, however, this problem is far from solved.

Thus, in creating a unified theory of Clebsch–Gordan coefficients of higher groups on the basis of the theory of angular momentum, the fact, noted comparatively long ago but unused for some time, that the Clebsch–Gordan coefficients of SU_2 can be expressed in terms of terminating hypergeometric series (${}_3F_2$ functions) with argument z = 1 [33, 34] as

$$\begin{pmatrix} j_1 & j_2 & j \\ m_1 & m_2 & m \end{pmatrix} = (-1)^{j_2+m_2} \frac{(j+j_2+m_1)!}{(j_1-j_2-m)!} \, {}_3F_2\begin{pmatrix} j_1-m_1+1, \; -j-m, \; -j+j_1-j_2 \\ j_1-j_2-m+1, \; -j-j_2-m_1 \end{pmatrix}; 1 \Big) \times$$
$$\times \left[\frac{(j_1+j-j_2)!\,(j_1+j_2-j)!\,(j-m)!\,(j_1-m_1)!\,(2j+1)}{(j_2-j_1+j)!\,(j_1+j_2+j+1)!\,(j+m)!\,(j_1+m_1)!\,(j_2-m_2)!\,(j_2+m_2)!}\right]^{1/2}, \tag{97}$$

where $\begin{pmatrix} j_1 & j_2 & j \\ m_1 & m_2 & m \end{pmatrix}$ denotes a Clebsch–Gordan coefficient in SU_2, is of great interest. Equation (97) gives just one of the possible ways of representing the Clebsch–Gordan coefficients of SU_2 in terms of an ${}_3F_2$ function. In all there are 24 different forms of this representation, not including those obtained by permutation of the parameters in ${}_3F_2\left(\begin{matrix} a_1, \, a_2, \, a_3 \\ b_1 b_2 \end{matrix}; 1\right)$ [33].

The fact that ${}_3F_2$ functions appear in the theory of angular momentum as established by Eq. (97) is not unique; the Racah coefficients $\begin{Bmatrix} j_1 & j_2 & j_3 \\ l_1 & l_2 & l_3 \end{Bmatrix}$ may be expressed in terms of a finite series in ${}_4F_3\left(\begin{matrix} \cdots \\ \cdots \end{matrix}; 1\right)$ of the particular form [33, 42]:

$$\begin{Bmatrix} j_1 & j_2 & j_3 \\ l_1 & l_2 & l_3 \end{Bmatrix} = \frac{\Delta(j_1 j_2 j_3)\,\Delta(j_1 l_2 l_3)\,\Delta(l_2 l_1 j_3)\,\Delta(j_2 l_1 l_3)}{(l_1+l_2-j_3)!\,(j_1+l_1-l_3)!\,(j_1+j_2-j_3)!\,(j_1+j_2-l_3)!} \frac{(-1)^{l_1+l_2+j_1+j_2}(j_1+j_2+l_1+l_2+1)!}{(l_3+j_3-l_2-j_2)!\,(l_3+j_3-l_1-j_1)!} \times$$
$$\times {}_4F_3\left(\begin{matrix} -j_1-j_2+j_3, \; -l_1-l_2+j_3, \; -j_1-l_2+l_3, \; -l_1-j_2+l_3 \\ -j_1-j_2-l_1-l_2-1, \; j_3+l_3-j_1-l_1+1, \; j_3+l_3-j_2-l_2+1 \end{matrix}; 1\right), \tag{98}$$

where

$$\Delta(abc) = \left[\frac{(a-b+c)!\,(a+b-c)!\,(-a+b+c)!}{(a+b+c+1)!}\right]^{1/2}.$$

This leads to the idea of expressing any 3nj-symbol in terms of ${}_pF_q$ functions (and, thus, translating the theory of angular momentum to the language of hypergeometric functions) since, according to [41, 42], arbitrary 3nj-symbols can be written in the form of weighted sums of products of Racah coefficients as follows: (a) expressions for 3nj-symbols of the first kind, $\begin{bmatrix} j_1 & j_2 \dots j_n \\ l_1 & l_2 \dots l_n \\ k_1 & k_2 \dots k_n \end{bmatrix}^1$:

$$\begin{bmatrix} j_1 & j_2 & \dots & j_n \\ l_1 & l_2 & \dots & l_n \\ k_1 & k_2 & \dots & k_n \end{bmatrix}^1 = \sum_x (2x+1)(-1)^{R+(n-1)x} \begin{Bmatrix} j_1 & k_1 & x \\ k_2 & j_2 & l_1 \end{Bmatrix} \begin{Bmatrix} j_2 & k_2 & x \\ k_3 & j_3 & l_2 \end{Bmatrix} \cdots \begin{Bmatrix} j_{n-1} & k_{n-1} & x \\ k_n & j_n & l_{n-1} \end{Bmatrix} \begin{Bmatrix} j_n & k_n & x \\ j_1 & k_1 & l_n \end{Bmatrix}; \tag{99}$$

and (b) expressions for 3nj-symbols of the second kind, $\begin{bmatrix} j_1 & j_2 \dots j_n \\ l_1 & l_2 \dots l_n \\ k_1 & k_2 \dots k_n \end{bmatrix}^2$:

$$\begin{bmatrix} j_1 & j_2 & \dots & j_n \\ l_1 & l_2 & \dots & l_n \\ k_1 & k_2 & \dots & k_n \end{bmatrix}^2 = \sum_x (2x+1)(-1)^{R+nx} \begin{Bmatrix} j_1 & k_1 & x \\ k_2 & j_2 & l_1 \end{Bmatrix} \begin{Bmatrix} j_2 & k_2 & x \\ k_3 & j_3 & l_2 \end{Bmatrix} \cdots \begin{Bmatrix} j_{n-1} & k_{n-1} & x \\ k_n & j_n & l_{n-1} \end{Bmatrix} \begin{Bmatrix} j_n & k_n & x \\ k_1 & j_1 & l_n \end{Bmatrix}, \tag{100}$$

where $R = \sum_{i=1}^{n} (j_i + l_i + k_i)$. However, it appears that already for the 9j-symbols there is no simple representation in terms of ${}_pF_q$ functions analogous to Eqs. (97) and (98) since, apparently, the simplest known numerical expression for the 9j-symbols [44, 45] contains (as opposed to the Clebsch–Gordan and Racah coefficients) triple sums of the form

$$\begin{Bmatrix} k & k_2 & k \\ l_1 & l_2 & l \\ j_1 & j_2 & j \end{Bmatrix} = \frac{(-1)^{j+l-k+j_2+l_2-k_2}\nabla(k_1k_2k)\,\nabla(l_2j_2k_2)\,\nabla(ljk)}{\nabla(k_1l_1j_1)\,\nabla(l_2l_1l)\,\nabla(jj_1j_2)} \times$$

$$\times \sum_{x,\,y,\,z} \frac{(-1)^{x+y+z}(k_1-l_1+j_1+x)!\,(2l_1-x)!\,(l-l_1+l_2+x)!\,(2j_2-y)!}{x!\,y!\,z!\,(j_1+l_1-k_1-x)!\,(l_1-l_2+l-x)!\,(k_1-l_1-j_2+j+x+y)!} \times$$

$$\times \frac{(j_1-j_2+j+y)!\,(k_1-k+l_2+j_2-y-z)!\,(k-k_1+k_2+z)!}{(l_2-l_1+k-j+x+z)!\,(j_2+l_2-k_2-y)!\,(k_2+l_2+j_2+1-y)!} \times$$

$$\times \frac{(k+l-j+z)!}{(j_1+j_2-j-y)!\,(j+l-k-z)!\,(k_1+k_2-k-z)!\,(2k+1+z)!}, \qquad (101)$$

where

$$\nabla(abc) = \left[\frac{(a-b+c)!\,(a+b-c)!\,(a+b+c+1)!}{(b+c-a)!}\right]^{1/2}.$$

This circumstance relates the 3nj-symbols ($n > 2$) to hypergeometric series having more than one argument (Appell functions and Horn and Laurincella series [37]).

It should be noted that the results listed in Eqs. (97) and (98) were found by analyzing expressions for the numerical values of the corresponding quantities obtained within the framework of the classical theory of angular momentum. For perfecting this theory, particularly for use as a basis for creating a unified theory of the Clebsch–Gordan coefficients of compact groups of physical symmetries, there will doubtless be interest in the systematic revision of the theory of angular momentum from the standpoint of the theory of hypergeometric functions. Without discussing the development of this idea in detail we note only that broad possibilities are opened up here in connection with the use of the CR representation of the ${}_pF_q$ functions and the expression of orthogonal polynomials in terms of them (see Section 2). We now demonstrate some of these possibilities using the results of Section 2.

Thus, Eq. (54b), when translated into the language of the theory of angular momentum with the aid of Eq. (97), shows that with nonstandard summation of the two weighted Clebsch–Gordan coefficients, some ${}_5F_4$ functions appear in Eq. (54). For special values of the parameters these ${}_5F_4$ functions degenerate into ${}_4F_3$ functions, or Racah coefficients, ${}_3F_2$ functions, or Clebsch–Gordan coefficients, or other quantities from the theory of angular momentum (see [35]). This makes it possible to introduce arbitrary ${}_pF_q$ functions into the theory of angular momentum as well as ${}_3F_2$ and ${}_4F_3$ functions.

An analogous translation of Eqs. (56) and (57) yields expressions for the Clebsch–Gordan coefficient in SU_2 in terms of a nonstandard sum of two [with Eq. (56)] or three [with Eq. (57)] weighted Clebsch–Gordan coefficients from SU_2:

$$\left(\begin{array}{c|c||c} j_1 & j_2 & j \\ m_1 & m_2 & m \end{array}\right) = (-1)^{j+m}\left[\frac{k!\,l!\,(j+m)!\,(2j+2j_2+j_1-m_1-k-l+1)}{(2j_2+j+m-2m_1)^{(k)}\,(j_1+m_1)^{(l)}}\right]^{1/2} \times$$

$$\times [2j+1]^{1/2} \sum_{\mu,\,\nu} \frac{(-1)^{\mu+\nu}}{(\mu+\nu-j-m)!\,(j+m-\nu)!\,(j+m-\mu)!} \times$$

$$\times \left[\frac{\mu!\,\nu!\,(2j+2j_2-\mu)^{(j+m-\mu)}\,(j_1+m_1-2j_2)^{(j+m-\nu)}}{(2j_2+2j-l)^{(\nu)}\,(j_1-m_1+\mu-k)^{(\mu)}\,(k+l-\mu)!} \times\right.$$

$$\times \left.\frac{(2j_2+j_1-m_1)^{(k-l+\nu-j-m)}}{(j-m+\nu+1)\,(2j+2j_2+j_1-m_1-k-l+\mu+1)}\right]^{1/2} \times$$

$$\times \left(\begin{array}{c|c||c} \tfrac{1}{2}(l+j_1-m_1) & \tfrac{1}{2}(l+k-\mu-j_1+m_1) & \tfrac{1}{2}(\mu-l-k+j_1-m_1)+j+j_2 \\ \tfrac{1}{2}(l-j_1+m_1) & \tfrac{1}{2}(k+\mu)-j-j_2 & \tfrac{1}{2}(\mu+l+k-j_1+m_1)-j-j_2 \end{array}\right) \times$$

$$\times \left(\begin{array}{l|l||l} j_1 - {}^1/_2 (l+k) & j_2 + {}^1/_2 (j+m-l-\nu) & j + {}^1/_2 (\nu - j - m) \\ m_1 + {}^1/_2 (k-l) & m_2 + {}^1/_2 (-j-m+l+\nu) & m + {}^1/_2 (\nu - j - m) \end{array}\right), \tag{102}$$

$$m = m_1 + m_2, \quad 0 \leqslant l \leqslant j_1 + m_1, \quad 0 \leqslant k \leqslant j + 2j_2 + m - 2m_1.$$

$$\left(\begin{array}{l|l||l} j_1 & j_2 & j \\ m_1 & m_2 & m_1+m_2 \end{array}\right) = \left(\begin{array}{l|l||l} {}^1/_2 (\beta + x) & {}^1/_2 (\beta+\alpha-p-q-x) & {}^1/_2 (b+p+q-\alpha-\beta) \\ {}^1/_2 (\beta - x) & {}^1/_2 (\alpha+p+q+x-b) & {}^1/_2 (p+q+\alpha+\beta-b) \end{array}\right) =$$
$$= \sum_{k,l,m} \left(\begin{array}{l|l||l} {}^1/_2 (\beta + x) & {}^1/_2 (\beta+\alpha-k-x) & {}^1/_2 (b-\beta-\alpha+k) \\ {}^1/_2 (\beta - x) & {}^1/_2 (\alpha+k+x-b) & {}^1/_2 (k+\beta+\alpha-b) \end{array}\right) \times$$
$$\times \left(\begin{array}{l|l||l} {}^1/_2 (\beta + x) & {}^1/_2 (\beta+\alpha-q-l-x) & {}^1/_2 (b+q+l-\alpha-\beta) \\ {}^1/_2 (\beta - x) & {}^1/_2 (\alpha+q+l+x-b) & {}^1/_2 (q+l+\alpha+\beta-b) \end{array}\right) \times$$
$$\times \left(\begin{array}{l|l||l} {}^1/_2 (a-\alpha-\beta) & {}^1/_2 (-\beta-m-x) & {}^1/_2 (b+m) \\ {}^1/_2 (\alpha-\beta-x) & {}^1/_2 (m-b+\beta+x-\alpha) & {}^1/_2 (m-b) \end{array}\right) \times$$
$$\times (-1)^{p+k+l+m} p^{(k)} \left[\frac{(b+1+q+p-\beta-\alpha)}{(b-\alpha-\beta+q+l+1)} \times\right.$$
$$\times \frac{(b-\alpha-\beta+1)\, m!\, k!\, \alpha!\, \beta!\, (-\alpha)^{(\beta+m)} (b-x-k)!}{(b+m+1)(p+q)^{(p-l)} (b-\beta-\alpha+k+1)(\alpha+\beta-k)!} \times$$
$$\times \left.\frac{(x+p+q-\alpha)^{(p-l)} (b-q-l-x)^{(p-l)} (\alpha+\beta-p-q)^{(l-p)} (b+1)}{(x+k-\alpha)^{(k)} (b-x)^{(\beta+m)} (x+m)^{(m)} (b-x-\beta)^{(b-x)}}\right]^{1/2} \times$$
$$\times \sum_n \frac{(p-n)!}{(k-n)!\,(m+l-p+n)!\,(p-l-n)!\,(p-m-n)!\,n!}. \tag{103}$$

These equations are recurrence relations for Clebsch–Gordan coefficients of quadratic and cubic type, respectively, and are not contained in either the classical variant of the theory of angular momentum or in its generalizations discussed in [33]. Without analyzing these formulas in detail, we note that Eq. (103) brings to mind the form of the relation between the Clebsch–Gordan and Racah coefficients in the classical theory of angular momentum. This allows us to treat the complicated factor in the sum with the Clebsch–Gordan coefficients as a generalized transformation matrix having roughly the significance of the Racah coefficients but numerically expressed in terms of an ${}_3F_2$ function.

Using Eqs. (97) and (64) it is possible to obtain an expression for the Clebsch–Gordan coefficients of SU_2 in terms of the orthogonal Hahn polynomials, $p_n(x; \beta, \gamma, \delta)$:

$$\left(\begin{array}{l|l||l} j_1 & j_2 & j \\ m_1 & m_2 & m \end{array}\right) = (-1)^{j-j_1-m_2} \times$$
$$\times \left[\frac{(j+j_1-j_2)!\,(j+j_2-j_1)!\,(j_1+j_2-j)!\,(j_1+m_1)!\,(j_1-m_1)!}{(j_1+j_2+j+1)!\,(j+m)!\,(j-m)!\,(j_2-m_2)!\,(j_2+m_2)!}\right]^{1/2} \times$$
$$\times [2j+1]^{1/2} p_{j+j_2-j_1} (j+m;\ j_1-j_2+1-m,\ -j-j_2-m_1,\ -m-j_2-j_1). \tag{104}$$

Since in certain cases the Hahn polynomials degenerate into discrete Chebyshev polynomials, $t_n(x, N)$, it is easy to find an expression for the latter in terms of Clebsch–Gordan coefficients of SU_2 of the particular form

$$t_n(x, N) \equiv p_n(x;\ 1,\ 1-N,\ 1-N) = \left[\frac{(N+n)^{(2n+1)}\, x!\, (x+N-1)^{(2x)}}{2n}\right]^{1/2} \left(\begin{array}{c|c||c} \frac{N-1}{2} & \frac{N-1}{2} & n \\ \frac{N-1}{2} - x & x + \frac{1-N}{2} & 0 \end{array}\right). \tag{105}$$

Thus, a close relationship has been established between the theory of angular momentum and the theory of orthogonal polynomials. This opens up the possibility of transferring the results and method of one theory to the other. In this respect it is interesting to note that the theory of angular momentum makes it possible to extend the classical list of systems of discrete orthogonal polynomials. Thus, keeping in mind the orthogonality of the Racah coefficients, it is possible to use Eq. (98) to verify that a system of orthogonal polynomials $\{\pi_s(x)\}$ of a single argument which is not one of the classical polynomials given in [37] can be expressed in terms of them. For example, making the parameter substitution

$$
\begin{gathered}
s=j_1+j_2-j_3, \quad F-x-\beta=l_1+l_2-j_3, \quad A-x=j_1+l_2-l_3, \\
B-x=l_1+j_2-l_3, \quad C+x=j_3+l_3-l_1-j_1+1, \quad \beta=j_1+j_2+j_3+1, \\
D+x=j_3+l_3-j_2-l_2+1, \quad F-x=j_1+j_2+l_1+l_2+1,
\end{gathered}
$$

in Eq. (98) it is possible to obtain a difference Rodriguez formula for $\pi_s(x)$ from Eq. (98) using the methods of Section 2 (to within some factor depending on the weighting function, j(x), which must also be defined):

$$
\pi_s(x) \simeq \begin{Bmatrix} j_1 & j_2 & j_3 \\ l_1 & l_2 & l_3 \end{Bmatrix} = (-1)^s N \frac{\Delta_x^s \{[(F-x)^{(\beta)} A^{(x)} B^{(x)}][(C+x)^{(x)} (D+x)^{(x)}]^{-1}\}}{[(F-x)^{(\beta)} A^{(x)} B^{(x)}][(C+x)^{(x)} (D+x)^{(x)}]^{-1}}, \tag{106}
$$

where N is the factor with the ${}_4F_3$ function in Eq. (98).

This result makes it possible to hope that the interrelationship between the theory of angular momentum and the theory of orthogonal polynomials will be fruitful for the development of both theories.

In concluding this section we note that the possibilities indicated here for developing the theory of angular momentum may be used in creating a unified theory of the Clebsch–Gordan coefficients of compact groups. One of the immediate goals may be taken to be (finding) an expression in the language of hypergeometric functions for the results from the algebraic theory of the Clebsch–Gordan coefficients of SU_n (see, for example, [34, 40]) with which the theory could be simplified and standardized using hypergeometric function techniques. A good premise for solving this problem is that expressions are known for the basis quantities of the theory of the Clebsch–Gordan coefficients of SU_n in terms of the Clebsch–Gordan coefficients of SU_2 [34, 40]. Here it is also interesting to note [46], where (using another approach) expressions are obtained for several sums of Clebsch–Gordan coefficients of the SU_n group in terms of ${}_3F_2$ functions and the Clebsch–Gordan coefficients of the SU_2 group which reduce to them.

4. Difference Methods in the Compact Group Formalism

In this final section we shall discuss several prospects for the application of difference techniques in the development of the compact group formalism, or more precisely, of a practically important area of it, the theory of Clebsch–Gordan coefficients.

The above results (the relationship between the theory of angular momentum and hypergeometric functions; the effeciveness of the CR representation in the theory of hypergeometric functions) make it possible to pose the problems of developing a finite difference formulation of the theory of angular momentum and of using it to create a unified theory of Clebsch–Gordan coefficients of compact groups. This would, on the one hand, make it possible to standardize the computational technique and, on the other, open up new, deeper interrelationships and symmetries among various quantities in different groups. In this regard it is noteworthy that difference expressions for the individual quantities in the symmetry formalism were published comparatively long ago. Thus, for example, in [42, 47, 48] expressions for the Clebsch–Gordan coefficients of SU_2 were written in a form which resembles the difference form [Eq. (62)] of the Rodriguez formulas for orthogonal polynomials. In [42] expressions are also given for several other quantities from the theory of angular momentum in terms of symbolic powers. Ansary has obtained an expression for the Clebsch–Gordan coefficients of SU_2 in the form of quasibinomials (expressions in terms of symbolic powers which formally resemble a binomial series) [49]. However, first of all, none of these results were combined into a system, and, second, they were not significantly utilized in the theory of Clebsch–Gordan coefficients. Meanwhile, the adequacy of discrete analysis as a formal language for the theory of Clebsch–Gordan coefficients lies at the very foundation of the latter.

As is known (see, for example, [50]), the theory of representations of continuous groups (including a branch of it, the theory of Clebsch–Gordan coefficients) may be developed in two ways: (1) by operating with the groups G themselves and functions $f(g)$ of the group elements g, where $g \in G$ (the integral approach), or (2) by using the Lie algebras of the groups (infinitesimal approach). In the first case all the results are expressed in terms of continuous analysis. In particular, the Clebsch–Gordan coefficients are determined by equations (see, for example, [33, 47, 48, 50]) containing matrix elements $T^{[j]}_{[m][m']}(g)$ of the finite transformation operators of an irreducible representation ($|j|$ is a composite signature of the irreducible representation and $[m] = [m_1, m_2, \ldots, m]$ is a set of numbers which enumerate the basis vectors of the irreducible representation). For compact groups this finally leads to an expression for the Clebsch–Gordan coefficients $\left(\begin{matrix}[j_1] \\ [m_1]\end{matrix}\middle|\begin{matrix}[j_2] \\ [m_2]\end{matrix}\middle\|\begin{matrix}[j] \\ [m]\end{matrix}\right]\gamma\Big)$ (where γ is an index which distinguishes multiple irreducible representation in the Clebsch–Gordan series [40]) with the aid of integrals over the group set of the matrix elements $T^{[j_1]}_{[m_1][m_1']}(g)$, $T^{[j_2]}_{[m_2][m_2']}(g)$, and $T^{[j]}_{[m][m']}(g)$. In the second case the results of the group formalism are expressed in terms of the Lie groups for infinitesimal group operators, the generators E_α and H_i (cf. Section 1). Then the Clebsch–Gordan coefficients are defined by a system of recurrence relations (difference equations) obtained when the generators operate on both types of basis related by the Clebsch–Gordan coefficients [40]:

$$\left|\begin{matrix}[j] \\ [m]\end{matrix}\right)_\gamma = \sum_{[m_1][m_2]} \left(\begin{matrix}[j_1] \\ [m_1]\end{matrix}\middle|\begin{matrix}[j_2] \\ [m_2]\end{matrix}\middle\|\begin{matrix}[j] \\ [m]\end{matrix}\right]\gamma\Big) \left|\begin{matrix}[j_1] \\ [m_1]\end{matrix}\right) \left|\begin{matrix}[j_2] \\ [m_2]\end{matrix}\right), \tag{107}$$

where $\left|\begin{matrix}[j_1] \\ [m_1]\end{matrix}\right)$, $\left|\begin{matrix}[j_2] \\ [m_2]\end{matrix}\right)$, $\left|\begin{matrix}[j] \\ [m]\end{matrix}\right)$ are the basis vectors of the corresponding irreducible representations. Nontrivial equations are obtained upon acting with the nondiagonal operators E_α and have the form (cf. Eq. (9.5) of [40]):

$$([m'] | E_\alpha([j]) | [m]) \left(\begin{matrix}[j_1] \\ [m_1]\end{matrix}\middle|\begin{matrix}[j_2] \\ [m_2]\end{matrix}\middle\|\begin{matrix}[j] \\ [m]\end{matrix}\right]\gamma\Big) =$$

$$= ([m_1] | E_\alpha([j_1]) | [m_1']) \left(\begin{matrix}[j_1] \\ [m_1']\end{matrix}\middle|\begin{matrix}[j_2] \\ [m_2]\end{matrix}\middle\|\begin{matrix}[j] \\ [m']\end{matrix}\right]\gamma\Big) + ([m_2] | E_\alpha([j_2]) | [m_2']) \left(\begin{matrix}[j_1] \\ [m_1]\end{matrix}\middle|\begin{matrix}[j_2] \\ [m_2']\end{matrix}\middle\|\begin{matrix}[j] \\ [m']\end{matrix}\right]\gamma\Big), \tag{108}$$

where $(\ldots | E_\alpha(\ldots) | \ldots)$ are matrix elements of the generators calculated in certain canonical bases of the spaces of the corresponding irreducible representations (explicit expressions for them in the SU_n group may be found in [50]) and [m'], etc. are sets of numbers which enumerate the basis vectors of the irreducible representation obtained from [m] by applying the difference operators $\Delta_i \pm 1$ for certain components m_i of [m]. Doubly repeated indices denote summation.

As for the interrelation between discrete and continuous analysis (cf. Section 2), it is interesting to note that acting on the basis vectors of the available differential generators (see Section 1) leads to difference equations for the Clebsch–Gordan coefficients. Comparing this fact with the fact that the bases of the ireducible representation and the matrix elements $T^{[j]}_{[m][m']}(g)$ are expressed in terms of special functions (of mathematical physics) of the group parameters (see [47]) as well as in terms of the previously derived expressions for the Clebsch–Gordan coefficients of SU_2 (in terms of ${}_3F_2$ functions with argument $z = 1$), we can advance the hypothesis (in the light of the comments of Section 2.2) that the Clebsch–Gordan coefficients of SU_n (and other compact groups) may be expressed in the form of discrete analogs of the functions, their generalizations, and combinations. However, to verify this hypothesis it will be necessary to make specific analyses and obtain explicit solutions for equations of the type (108) for various groups. In doing this it would be convenient to exploit as fully as possible the parallels with the method of continuous analysis in the theory of special functions. In particular, the system of defining equations (108), which are, in general, difference analogs of partial differential equations, may be reduced, for this analysis, to a single equation of higher orders in the operators Δ_i. Then, using the factorizability of the Clebsch–Gordan coefficients, it is possible (as in the case of the continuous problems of mathematical physics) to use the method of separation of variables and obtain special difference equations for the G_i scalar factors (cf. Section 3). When the above hypotheses are true, it might be expected that the number of such special equations (by analogy with continuous analysis) would be comparatively small and that they would be classified by groups (cf. the classification of "continuous" special functions in [47]). This latter aspect is of special interest from the standpoint of analyzing and solving the difference equations which arise in the theory of coherent-cooperative phenomena and other physical problems.

Another aspect of applying difference methods in the compact group formalism has to do with the generating functions $\Psi(x)$ for the ${}_pF_q$ functions and their generalization (see Section 2). Using the CR representations or the generating functions for the fundamental quantities of the theory, we can use the method of Section 2 to obtain CR representations (or generating functions) and then numerical expressions for arbitrary quantities (algebraic combinations of the fundamental quantities). Since the theory of angular momentum occupies a central place in the construction of a unified theory of Clebsch–Gordan coefficients of compact groups (cf. Section 3), it is appropriate to derive generating functions or CR representations for the fundamental quantities of the theory of angular momentum. The CR representation for the Clebsch–Gordan and Racah coefficients is given by Eqs. (50d), (97), and (106). In addition to these we now write the CR representations (special form) for the Ansary quasibinomial and the 9j sym-

$$((y\eta - z\varsigma))^{(s)} = (-1)^s \Delta_x^s \left[z^{(x)} \varsigma^{(x)} y^{(s-x)} \eta^{(s-x)} \right]\Big|_{x=0}, \tag{109a}$$

$$\begin{Bmatrix} k_1 & k_2 & k_3 \\ l_1 & l_2 & l_3 \\ j_1 & j_2 & j_3 \end{Bmatrix} = \frac{(-1)^{j_1+l_1-k_1} (j_1 - j_2 + j)!\,(k + l - j)!}{(j_1 + l_1 - k_1)!\,(j_2 + l_2 - k_2)!\,(j + l - k)!} \times$$

$$\times \frac{\nabla(k_1 k_2 k)\,\nabla.(l_2 j_2 k_2)\,\nabla(l j k)}{\nabla(k_1 l_1 j_1)\,\nabla(l_2 l_1 l)\,\nabla(j j_1 j_2)\,(j_2 + l_2 + k_2 + 1)!}\, \Delta_u^{j_1+l_1-k_1} \Delta_v^{j_2+l_2-k_2} \Delta_w^{j+l-k} \left[\frac{(k_1 - l_1 + j_1 + u)^{(j_1+j_2-j-v)}}{(l_1 - l_2 + l - u)^{(l-l_1-l_2)}} \times \right.$$

$$\left. \times \frac{(l - l_1 + l_2 + u)^{(l+j-k-w)} (j_1 - j_2 + j + v)^{(v)} (k + l - j + w)^{(w)}}{(2k + 1 + w)^{(k_1+k-k_2+1)} (k_1 + k_2 - k - w)^{(k_2-l_2-j_2+v)}} \frac{(k_2 + l_2 + j_2 + 1)^{(v)}}{(j_1 + j_2 - j - v)^{(j_1-j-j_2)}} \right]\Bigg|_{\substack{u=0\\v=0\\w=0}}. \tag{109b}$$

These equations are easily derived (by the methods of Section 2) from Eqs. (4.1) of [35] and Eq. (101) of the present paper, respectively.

With this we conclude a very short discussion of the prospects for using difference methods to develop the compact group formalism and note that with the aid of the technique of analytic continuation applied in the published literature on physics (see several examples of this type for the SU (1,1) and SL (2C) groups in [45, 51, and 52]) it should be possible to partially extend these results to noncompact groups as well. In doing this the comments in Section 2.2 about extending the CR representation of finite (terminating) hypergeometric series to arbitrary ${}_pF_q$ functions may be useful.

In conclusion we point out several possibilities in the theory of coherent phenomena which flow from the results of our investigation. Here we note the following. While the relation between coherent phenomena and nonlinear differential equations has recently begun to draw considerable attention, the finite-difference approach is practically unused. Here we have pointed out the naturalness of difference and group theoretical treatments of certain quantities from the theory of coherent phenomena, as well as the unity and parallelism of the continuous and discrete aspects of all the mathematical questions considered here (which evidently has a fundamental, although not yet fully understood, significance). Thus, the results derived here open up a particular approach for the finite difference analysis of coherent phenomena. It remains to construct specific difference formulations of the theory of coherent cooperative phenomena and to analyze their physical features.

APPENDIX 1. THE CARTAN MATRICES

$$A_l\,(SU_{l+1}):\quad A_{ij}=\left\|\begin{array}{cccccc} 2 & -1 & 0\dots & 0 & 0 \\ -1 & 2 & -1\dots & 0 & 0 \\ 0 & -1 & 2\dots & 0 & 0 \\ \cdot & \cdot & \cdot & \cdot & \cdot \\ 0 & 0 & 0\dots & 2 & -1 \\ 0 & 0 & 0\dots & -1 & 2 \end{array}\right\|$$

$$B_l:\quad A_{ij}=\left\|\begin{array}{cccccc} 2 & -1 & 0\dots & 0 & 0 & 0 \\ -1 & 2 & -1\dots & 0 & 0 & 0 \\ 0 & -1 & 2\dots & 0 & 0 & 0 \\ \cdot & \cdot & \cdot & \cdot & \cdot & \cdot \\ 0 & 0 & 0\dots & 2 & -1 & 0 \\ 0 & 0 & 0\dots & -1 & 2 & -1 \\ 0 & 0 & 0\dots & 0 & -2 & 2 \end{array}\right\|$$

$$C_l:\quad A_{ij}=\left\|\begin{array}{cccccc} 2 & -1 & 0\dots & 0 & 0 & 0 \\ -1 & 2 & -1\dots & 0 & 0 & 0 \\ 0 & -1 & 2\dots & 0 & 0 & 0 \\ \cdot & \cdot & \cdot & \cdot & \cdot & \cdot \\ 0 & 0 & 0\dots & 2 & -1 & 0 \\ 0 & 0 & 0\dots & -1 & 2 & -2 \\ 0 & 0 & 0\dots & 0 & -1 & 2 \end{array}\right\|$$

$$D_l:\quad A_{ij}=\left\|\begin{array}{cccccc} 2 & -1 & 0\dots & 0 & 0 & 0 \\ -1 & 2 & -1\dots & 0 & 0 & 0 \\ 0 & -1 & 2\dots & 0 & 0 & 0 \\ \cdot & \cdot & \cdot & \cdot & \cdot & \cdot \\ 0 & 0 & 0\dots & 2 & -1 & -1 \\ 0 & 0 & 0\dots & -1 & 2 & 0 \\ 0 & 0 & 0\dots & -1 & 0 & 2 \end{array}\right\|$$

$$G_2:\quad A_{ij}=\left\|\begin{array}{cc} 2 & -1 \\ -3 & 2 \end{array}\right\| \qquad F_4:\quad A_{ij}=\left\|\begin{array}{cccc} 2 & -1 & 0 & 0 \\ -1 & 2 & -2 & 0 \\ 0 & -1 & 2 & -1 \\ 0 & 0 & -1 & 2 \end{array}\right\|$$

$$E_6:\quad A_{ij}=\left\|\begin{array}{cccccc} 2 & -1 & 0 & 0 & 0 & 0 \\ -1 & 2 & -1 & 0 & 0 & 0 \\ 0 & -1 & 2 & -1 & 0 & -1 \\ 0 & 0 & -1 & 2 & -1 & 0 \\ 0 & 0 & 0 & -1 & 2 & 0 \\ 0 & 0 & -1 & 0 & 0 & 2 \end{array}\right\|$$

$$E_7:\quad A_{ij}=\left\|\begin{array}{ccccccc} 2 & -1 & 0 & 0 & 0 & 0 & 0 \\ -1 & 2 & -1 & 0 & 0 & 0 & 0 \\ 0 & -1 & 2 & -1 & 0 & 0 & 0 \\ 0 & 0 & -1 & 2 & -1 & 0 & -1 \\ 0 & 0 & 0 & -1 & 2 & -1 & 0 \\ 0 & 0 & 0 & 0 & -1 & 2 & 0 \\ 0 & 0 & 0 & -1 & 0 & 0 & 2 \end{array}\right\|$$

APPENDIX 2. BASIS ROOTS

A_2: $\alpha_1=\{1, 0\}$

$\alpha_2=\left\{-\frac{1}{2}, \frac{\sqrt{3}}{2}\right\}$

A_3: $\alpha_1=\{1, 0, 0\}$

$\alpha_2=\left\{-\frac{1}{2}, \frac{\sqrt{3}}{2}, 0\right\}$

$\alpha_3=\left\{0, -\frac{1}{\sqrt{3}}, \frac{\sqrt{2}}{\sqrt{3}}\right\}$

A_4: $\alpha_1=\{1, 0, 0, 0\}$

$\alpha_2=\left\{-\frac{1}{2}, \frac{\sqrt{3}}{2}, 0, 0\right\}$

$\alpha_3=\left\{0, -\frac{1}{\sqrt{3}}, \frac{\sqrt{2}}{\sqrt{3}}, 0\right\}$

$\alpha_4=\left\{0, 0, -\frac{\sqrt{3}}{2\sqrt{2}}, \frac{\sqrt{5}}{2\sqrt{2}}\right\}$

A_5: $\alpha_1=\{1, 0, 0, 0, 0\}$

$\alpha_2=\left\{-\frac{1}{2}, \frac{\sqrt{3}}{2}, 0, 0, 0\right\}$

$\alpha_3=\left\{0, -\frac{1}{\sqrt{3}}, \frac{\sqrt{2}}{\sqrt{3}}, 0, 0\right\}$

$\alpha_4=\left\{0, 0, -\frac{\sqrt{3}}{2\sqrt{2}}, \frac{\sqrt{5}}{2\sqrt{2}}, 0\right\}$

$\alpha_5=\left\{0, 0, 0, -\frac{\sqrt{2}}{\sqrt{5}}, \frac{\sqrt{3}}{\sqrt{5}}\right\}$

B_2: $\alpha_1=\{\sqrt{2}, 0\}$

$\alpha_2=\left\{-\frac{1}{\sqrt{2}}, \frac{1}{\sqrt{2}}\right\}$

B_3: $\alpha_1=\{\sqrt{2}, 0, 0\}$

$\alpha_2=\left\{-\frac{1}{\sqrt{2}}, \frac{\sqrt{3}}{2}, 0\right\}$

$\alpha_3=\left\{0, -\frac{\sqrt{2}}{\sqrt{3}}, \frac{1}{\sqrt{3}}\right\}$

B_4: $\alpha_1=\{\sqrt{2}, 0, 0, 0\}$

$\alpha_2=\left\{-\frac{1}{\sqrt{2}}, \frac{\sqrt{3}}{2}, 0, 0\right\}$

$\alpha_3=\left\{0, -\frac{\sqrt{2}}{3}, \frac{2}{\sqrt{3}}, 0\right\}$

$\alpha_4=\left\{0, 0, -\frac{\sqrt{3}}{2}, \frac{1}{2}\right\}$

C_2: $\alpha_1=\{1, 0\}$

$\alpha_2=\{-1, 1\}$

C_3: $\alpha_1=\{1, 0, 0\}$

$\alpha_2=\left\{-\frac{1}{2}, \frac{\sqrt{3}}{2}, 0\right\}$

$\alpha_3=\left\{0, -\frac{2}{\sqrt{3}}, \frac{\sqrt{2}}{\sqrt{3}}\right\}$

C_4: $\alpha_1=\{1, 0, 0, 0\}$

$\alpha_2=\left\{-\frac{1}{2}, \frac{\sqrt{3}}{2}, 0, 0\right\}$

$\alpha_3=\left\{0, -\frac{1}{\sqrt{3}}, \frac{\sqrt{2}}{\sqrt{3}}, 0\right\}$

$\alpha_4=\left\{0, 0, -\frac{\sqrt{3}}{2}, \frac{1}{\sqrt{2}}\right\}$

D_4: $\alpha_1=\{1, 0, 0, 0\}$

$\alpha_2=\left\{-\frac{1}{2}, \frac{1}{2}, -\frac{1}{2}, -\frac{1}{2}\right\}$

$\alpha_3=\{0, 0, 1, 0\}$

$\alpha_4=\{0, 0, 0, 1\}$

D_5: $\alpha_1=\{1, 0, 0, 0, 0\}$

$\alpha_2=\left\{-\frac{1}{2}, \frac{\sqrt{3}}{2}, 0, 0, 0\right\}$

$\alpha_3=\left\{0, -\frac{1}{\sqrt{3}}, \frac{1}{\sqrt{6}}, -\frac{1}{2}, -\frac{1}{2}\right\}$

$\alpha_4=\{0, 0, 0, 1, 0\}$

$\alpha_5=\{0, 0, 0, 0, 1\}$

G_2: $\alpha_1=\{1, 0\}$

$\alpha_2=\left\{-\frac{2}{3}, \frac{5}{9}\right\}$

E_6: $\alpha_1=\{1, 0, 0, 0, 0, 0\}$

$\alpha_2=\left\{-\frac{1}{2}, \frac{\sqrt{3}}{2}, 0, 0, 0, 0\right\}$

F_4: $\alpha_1 = \{1,\ 0,\ 0,\ 0\}$

$\alpha_2 = \left\{-\frac{1}{2},\ \frac{\sqrt{3}}{2},\ 0,\ 0\right\}$

$\alpha_3 = \left\{0,\ -\frac{2}{\sqrt{3}},\ \frac{\sqrt{2}}{\sqrt{3}},\ 0\right\}$

$\alpha_4 = \left\{0,\ 0,\ -\frac{\sqrt{3}}{\sqrt{2}},\ \frac{1}{\sqrt{2}}\right\}$

$\alpha_3 = \left\{0,\ -\frac{1}{\sqrt{3}},\ \frac{\sqrt{5}}{2\sqrt{3}},\ 0,\ 0,\ -\frac{1}{2}\right\}$

$\alpha_4 = \left\{0,\ 0,\ -\frac{\sqrt{3}}{\sqrt{5}},\ \frac{\sqrt{2}}{\sqrt{5}},\ 0,\ 0\right\}$

$\alpha_5 = \left\{0,\ 0,\ 0,\ -\frac{\sqrt{5}}{2\sqrt{2}},\ \frac{\sqrt{3}}{2\sqrt{2}},\ 0\right\}$

$\alpha_6 = \{0,\ 0,\ 0,\ 0,\ 0,\ 1\}$

LITERATURE CITED

1. R. H. Dicke, Phys. Rev., 93:99 (1954).
2. L. A. Shelepin, Zh. Éksp. Teor. Fiz., 54:1463 (1968).
3. É. Sh. Teplitskii, Teor. Mat. Fiz., 2:399 (1970).
4. T. M. Makhviladze and L. A. Shelepin, Zh. Éksp. Teor. Fiz., 62:2066 (1972).
5. V. M. Akulin, S. S. Alimpiev, N. V. Karlov, and L. A. Shelepin, Trudy FIAN, 87:141 (1976) (present volume).
6. R. Glauber, in: Quantum Optics, Academic Press (1969).
7. D. Klauder and E. Sudarshan, Fundamentals of Quantum Optics, Benjamin (1968).
8. R. Bonifacio, P. Schwendimann, and F. Haake, Phys. Rev., A4:302,854 (1971).
9. R. Bonifacio, D. M. Kim, and M. O. Scully, Phys. Rev., 187:441 (1969).
10. P. W. Atkins and I. C. Dobson, Proc. Roy. Soc. London, A321:371 (1971).
11. T. M. Makhviladze and L. A. Shelepin, Yad. Fiz., 15:1081 (1972).
12. B. M. Levitan, The Theory of Generalized Displacement Operators [in Russian], Nauka, Moscow (1973).
13. V. M. Fain, Photons and Nonlinear Media [in Russian], Sov. Radio, Moscow (1972).
14. I. A. Malkin and V. I. Man'ko, FIAN Preprint No. 15 (1971).
15. A. O. Barut and L. Girardello, Commun. Math. Phys., 21:41 (1971).
16. A. M. Perelomov, Commun. Math. Phys., 26:222 (1972).
17. B. Judd, Second Quantization and Atomic Spectroscopy, Johns Hopkins Press (1967).
18. B. Judd and B. Wayborn, The Theory of Complex Atomic Spectra [Russian translation], Mir, Moscow (1973).
19. N. Jacobson, Lie Algebras [Russian translation], Mir, Moscow (1964).
20. T. M. Makhviladze and L. A. Shelepin, Yad. Fiz., 15:608 (1972).
21. R. M. Asherova and Yu. F. Smirnov, Nucl. Phys., B4:399 (1968); A144:116 (1970).
22. F. I. Fedorov, Zh. Éksp. Teor. Fiz., 35:493 (1958).
23. E. E. Tkharev and F. I. Fedorov, Yad. Fiz., 5:1112 (1967).
24. L. A. Shelepin, Trudy FIAN, 30:253 (1964).
25. I. M. Lizin and L. A. Shelepin, Yad. Fiz., 9:440 (1969).
26. V. L. Ginzburg and L. D. Landau, Zh. Éksp. Teor. Fiz., 20:1064 (1950).
27. A. O. Gel'fond, The Calculus of Finite Differences [in Russian], Nauka, Moscow (1967).
28. L. M. Milne-Thomson, The Calculus of Finite Differences, Cambridge University Press (1951).
29. R. W. Hamming, Numerical Methods for Scientists and Engineers, McGraw-Hill (1962).
30. A. A. Samarskii, Introduction to the Theory of Difference Schemes [in Russian], Nauka, Moscow (1971); S. K. Godunov and V. S. Ryaben'kii, Difference Schemes. An Introduction to the Theory [in Russian], Nauka, Moscow (1973).
31. R. Richtmeyer and K. Morton, Difference Methods for Initial-Value Problems, Wiley (1967).
32. V. G. Kadyshevskii, R. M. Mir-Kasimov, and M. Friman, Yad. Fiz., 9:646 (1969); V. G. Kadyshevskii (Kadyshevsky) and M. D. Matveev, Nuovo Cimento, 55A:275 (1968); V. G. Kadyshevskii, R. M. Mir-Kasimov, and N. B. Skachkov, Preprint Joint Institute for Nuclear Research E-2-3949, Dubna (1968).

33. Ya. A. Smorodinskii and L. A. Shelepin, Usp. Fiz. Nauk, 106:3 (1972).
34. L. A. Shelepin, Trudy FIAN, 70:3 (1973).
35. V. P. Karasev and L. A. Shelepin, Teor. Mat. Fiz., 17:67 (1973).
36. O. Madelung, Mathematical Methods of Physics [Russian translation], Fizmatgiz, Moscow (1961).
37. H. Bateman and A. Erdelyi, Higher Transcendental Functions [Russian translation], Vols. 1 and 2, Nauka, Moscow (1973, 1974).
38. L. J. Slater, Generalized Hypergeometric Functions, Cambridge University. Press (1966).
39. M. Hamermesh, Group Theory and Its Application to Physical Problems, Addison-Wisley (1962).
40. V. P. Karasev, Trudy FIAN, 70:147 (1973).
41. A. P. Yutsis, (Jucys), I. B. Levinson, and V. V. Vanagas, Mathematical Methods in the Theory of Angular Momentum [in Russian], GIPNL Lit. SSR, Vilnyus (1960).
42. A. P. Yutsis (Jucys) and A. A. Bandzitis, The Theory of Angular Momentum in Quantum Mechanics [in Russian], Mintis, Vilnyus (1965).
44. S. I. Ališauskas and A. P. Jucys, J. Math. Phys., 12:594 (1971); 13:575 (1972).
45. S. I. Alishauskas (Ališauskas), Litov. Fiz. Sb., 13:831 (1973).
46. G. I. Kuznetsov, Teor. Mat. Fiz., 18:358 (1974).
47. N. Ya. Vilenkin, Special Functions and the Theory of Group Representations [in Russian], Nauka, Moscow (1965).
48. I. I. Gel'fand, R. A. Minlos, and Z. Ya. Shapiro, Representations of Rotation Groups and the Lorentz Group [in Russian], Fizmatgiz, Moscow (1958).
49. S. M. Ansary, J. Math. Phys., 9:1299 (1968); Fortschr. Phys., 15:707, 729 (1967).
50. D. P. Zhelobenko, Compact Lie Groups and Their Representations [in Russian], Nauka, Moscow (1970).
51. R. L. Anderson, R. Raczka, M. A. Rashid, and P. Winternitz, J. Math. Phys., 11:1050 (1970).
52. A. D'Adda, R. D'Auria, and G. Ponzano, Generalized 6j-Coefficients and Their Symmetries. Preprint, University of Torino (Turin), June, 1972.

A STUDY OF THE COHERENT INTERACTION OF PULSED INFRARED RADIATION WITH MOLECULAR GASES†

S. S. Alimpiev

The coherent interactions of CO_2 laser light with the molecular gases BCl_3 and SF_6 are studied experimentally. Such characteristics of these gases as the homogeneous widths of absorption transition lines, transition dipole moments, the relative number of particles involved in absorption, and so on, are found. Some conclusions are drawn about the types of transitions studied.

INTRODUCTION

The study of the nonlinear resonant interaction of a field with matter has always been of great interest since this interaction serves as a powerful instrument for the investigation of many of the most important properties of matter. It is sufficient to recall such stages in the development of quantum electronics as electron paramagnetic resonance and nuclear magnetic resonance. The development of sources of coherent optical radiation has made it possible to carry the study of resonant interaction processes into the optical range.

The nonlinear resonant interaction of a field with matter leads to the most interesting effects when the interaction is coherent. The coherent nature of the interaction appears when the characteristic time of the interaction between the field and the matter τ_{int} is much less than the relaxation time T_2 of the polarization induced in the medium by the field (usually known as the tranverse or spin–spin relaxation time), that is, when

$$\tau_{int} \ll T_2. \tag{1}$$

When this condition is fulfilled such coherent interactions between a light field and matter as self-induced transparency, the photon echo, and nutation are observed.

In the case of self-induced transparency the interaction time of the field with matter is the duration of the radiation pulse propagating in the resonant medium, and in the case of the photon echo it is the time interval between the first exciting pulse and the photon echo pulse. The characteristic interaction time for the nutation effect is the nutation period.

The relaxation time T_2 for gases is determined by molecular collisions and, therefore, is a function of the particle concentration. The time T_2 increases as the gas pressure is re-

†This article consists of material from a dissertation submitted for the degree of Candidate of Physical and Mathematical Sciences defended at the Physics Institute of the Academy of Sciences in 1973. The research adviser was Prof. N. V. Karlov.

duced; hence, coherent interaction effects are usually observed at pressures very much less than atmospheric although the use of pulses of duration 10^{-10}-10^{-11} sec makes it possible to observe these effects at atmospheric pressure.

The coherent interaction effects produced by the interaction of an electromagnetic light field with matter are essentially optical analogs of the already well-known nuclear magnetic resonance spin echo and nutation effects which arise in an ensemble of nuclear spins situated in a constant magnetic field when a weak variable magnetic field at a resonant frequency is applied. The pioneering papers on this topic are those by Block [1] and Hahn [2].

It became possible to carry over the effects of coherent interactions to the optical range only with the invention of coherent radiation sources, that is, lasers. The photon echo effect was first observed in ruby with the aid of a ruby laser by Abella, Kurnit, and Hartmann [3] in 1964. McCall and Hahn obtained self-induced transparency, also in ruby [4]. In 1967 Patel and Slusher first obtained self-induced transparency and photon echo in gaseous SF_6 in the infrared (CO_2 laser light) [5, 6].

Somewhat later Hocker and Tang also observed nutation in a CO_2 laser–SF_6 system [7]. Basically these papers reported the observation of the effects involved. In addition, these effects are a very convenient instrument for studying many of the most important characteristics of absorbing media. Carrying coherent interaction effects over to the infrared opens up a new class of resonant objects, the molecular gases, which require detailed study. The present paper is devoted to an investigation of coherent effects between CO_2 laser light and two molecular gases, BCl_3 and SF_6.

These gases have already gained widespread application along with CO_2 lasers for passive Q-switching, laser frequency tuning, mode synchronization (mode locking) and frequency stabilization in laser chemistry research. While SF_6 has been used in quantum electronics for a relatively long time, gaseous boron trichloride has been discovered for this purpose only recently [8] and thus has been studied less.

The boron in the BCl_3 molecule appears in nature in a mixture of two stable iostopes B^{10}(19%) and B^{11} (81%). This results in the appearance of two vibrational levels ν_3 with energies 956 cm^{-1} and 995 cm^{-1} for $B^{11}Cl_3$ and $B^{10}Cl_3$, respectively. Thus, practically all the CO_2 laser spectrum is covered, with the less active laser lines of the R branch corresponding to the rarer isotope B^{10}. This peculiarity of the absorption spectrum of BCl_3 will doubtless assure it wide application in CO_2 laser technology in the future.

The interest in the molecular gases SF_6 and, especially, BCl_3 as new materials for quantum electronics has also stimulated our work. Our goal has been to determine the important features of these materials (the polarization relaxation time T_2, the magnitude of the matrix element of the dipole moment of the absorbing transitions of the molecule, etc.), knowledge of which is necessary for their more complete utilization with CO_2 lasers. It should be kept in mind that these parameters of molecular gases can only be measured by the methods of nonlinear spectroscopy, one of the branches of which is coherent interaction effects. Thus, for example, with the best apparatus for ordinary nonlaser spectroscopy it is not possible even to resolve the rotational structure of the vibrational bands of the molecules we are studying. The building of a parametric oscillator with a tunable range of 5-15 μm in the near future will still further increase the effectiveness of nonlinear spectroscopic methods and will also greatly expand the class of objects which can be studied. The development of an experimental technique and an analysis of the methods for observing and studying coherent interaction effects in CO_2 laser–BCl_3 and SF_6 systems will certainly make further research in this area easier.

Chapter I of this paper is devoted to a study of self-induced transparency in gaseous BCl_3. This effect was first observed by McCall and Hahn in ruby when a pulse of ruby laser light was passed through it. These authors have also made the most comprehensive theoretical description of this effect [4]. In the infrared at the CO_2 laser wavelength self-induced transparency was observed by Patel and Slusher in gaseous SF_6 [5]; however, failure to include the degeneracy (which always occurs in molecular gases) of the absorbing transition makes Patel's results somewhat incomplete. Rhodes and Javan have studied self-induced transparency taking this degeneracy into account in a CO_2 laser$-SF_6$ system [10].

Self-induced transparency is observed when a pulse of monochromatic light of duration satisfying the fundamental condition for a coherent interaction, $\tau_{pulse} \ll T_2$, propagates in a resonantly absorbing medium. The characteristic parameter of the pulse$-$medium system is θ, which is proportional to the area under the pulse and to the magnitude of the matrix element of the dipole moment of the transition μ; that is, $\theta(z) = \frac{\mu}{\hbar} \int\limits_{-\infty}^{\infty} \mathcal{E}(z, t)\, dt$, which equals $\mu\mathcal{E}\tau_{pulse}/\hbar$ for a rectangular pulse where $\mathcal{E}(z, t)$ is the slowly varying (compared to the optical frequency) amplitude of the pulse field, $\mathbf{E} = \mathbf{x}\mathcal{E}(z, t) \cos(\omega t - kz)$.

At low intensities ($\theta \ll 1$) the pulse is absorbed linearly by the medium without distortion according to the Bouguer$-$Beer law.

A pulse with values of $\mathcal{E}$ and τ_{pulse} at the entrance to the medium such that $\theta = (\mu/\hbar)\mathcal{E} \times \tau_{pulse} = \pi$ (such a pulse is called a π-pulse) completely inverts the system. If θ is slightly greater than π, then due to the action of the remaining part of the pulse the system undergoes stimulated reemission in the trailing edge of the pulse. Depending on how far it propagates in the medium the pulse becomes symmetric, is broadened, and is delayed due to absorption by the medium of its leading edge and reemission by the medium in the trailing edge. Then the value of θ approaches 2π. The pulse propagates further without distortion or loss of energy with a constant velocity $v < c$ since that part of the energy lost in the first half of the pulse due to absorption and, therefore, to excitation of the system, is exactly compensated by an equal amount of energy radiated by the system during the second half. If θ is somewhat less than π, then the system, although it remains practically completely inverted after such a pulse passes, reradiates spontaneously; that is, it radiates very little, since the radiative lifetime of the vibrational$-$rotational transitions of molecular gases is of the order of seconds and any kind of cooperative effects will be destroyed in this case due to phase randomization caused by inhomogeneous broadening.

Therefore, a pulse with a parameter somewhat less than π is dampened as it propagates in the medium, and the value of θ approaches zero.

In a nondegenerate system 3π-, 5π-, etc., pulses behave similarly to a π-pulse. However, taking degeneracy, which always occurs in real systems (in our case, molecular gases, this is a spatial degeneracy of vibrational$-$rotational transitions in the projection of the angular momentum in the direction of the pulse field), into account distorts the effect. However, when $\theta < 2\pi$ and the optical thickness of the system is of the order of several damping lengths, a degenerate system behaves similarly to a nondegenerate one. We then can estimate the matrix element of the dipole moment of the transition using the relation $(\mu/\hbar)\tau_{pulse}\mathcal{E} \approx \pi$ once we have isolated the range of pulse durations and intensities such that the pulses behave as π-pulses in a nondegenerate system. In addition, we can estimate the time T_2 from the way the pulse shape varies when the condition of Eq. (1) is artificially violated.

We used a pulsed CO_2 laser with a diffraction grating in its cavity for tuning over the emission lines as a light source for observing self-induced transparency in BCl_3. A rotating mirror Q-switch was used. The pulse lengths were τ_{pulse} = 100-300 nsec and the maximum intensity was several kW/cm^2. The test gas was put in a 3-mm-long cell with vacuum windows of NaCl. The radiation was picked up by two Ge$-$Zn$-$Sb infrared detectors located at the input and output ends of the cell.

The dependence of the intensity, shape, and delay of the output pulse on the intensity of the input pulse was studied experimentally. This made it possible to evaluate μ and T_2 in BCl_3 and to experimentally demonstrate the substantial effect of degeneracies on self-induced transparency in BCl_3.

The second chapter deals with a study of the photon echo in the molecular gases BCl_3 and SF_6. This effect, one of the most interesting coherent interactions, is the optical analog of the spin echo effect first studied by Hahn in NMR (nuclear magnetic resonance) [2]. Many theoretical papers are devoted to a description of the photon echo. Of these we should point out the work by Gordon et al. [11], in which the photon echo effect is examined in gases and degeneracy is included. Several simplifying assumptions were made there which limit the discussion to an optically thin medium; that is, the second-order effect of polarization on the field is neglected. In addition, inhomogeneous broadening at the moment the exciting pulse is acting is not included. The photon echo problem has been solved under much wider assumptions in a series of papers by A. I. Alekseev and I. V. Evseev [12]; however, they restricted themselves to the case of small values of the angular momentum J. The work by Kurnit et al. [9] dealing with the photon echo from the standpoint of the Dicke theory [13] should also be noted.

The photon echo arises when a medium is successively illuminated by two exciting pulses at a frequency equal to the frequency of an absorbing transition of the medium. Inhomogeneous (Doppler broadening in the case of a gaseous medium) broadening of the transition is then important.

In treating this effect it is convenient to use the vector model proposed by Hahn (see Fig. 1). The state of the system is described by a vector **R** in a coordinate system rotating with the

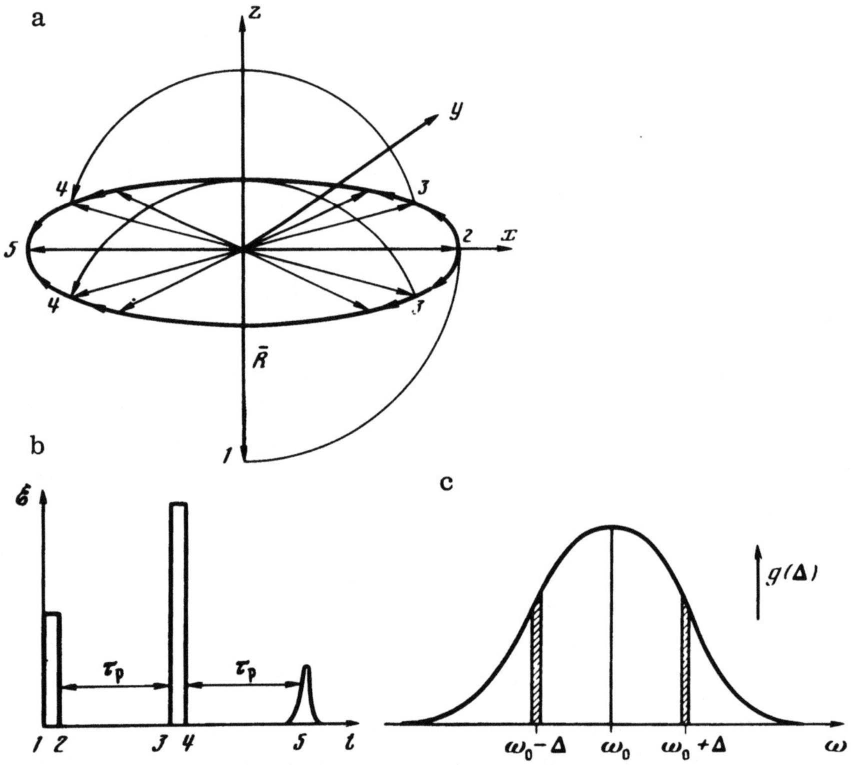

Fig. 1. The vector model for explaining the formation of a photon echo (a) of an inhomogeneously broadened absorbing line $g(\Delta)$ (c) due to the action of two perturbing $\pi/2$ and π pulses of radiation at frequency $\omega = \omega_0$ (b) separated by a time interval τ_p.

frequency of the field about the z axis. The transverse components of this vector are proportional to the absorption and dispersion components of the polarization of the medium, and the projection on the z axis is proportional to the difference in the populations of the system. Until the arrival of the exciting pulses, **R** is directed along the z axis (the medium is unpolarized and all the particles are in the lower level). The first exciting pulse is a $\pi/2$ pulse. As is known, a $\pi/2$ pulse produces maximal polarization of the medium and equalizes the populations of the upper and lower levels; that is, it rotates the vector **R** by $\pi/2$. This kind of ensemble state is known as superradiant (a Dicke ensemble [13]). The radiative lifetime of this ensemble due to spontaneous emission T_s is much less than the radiative lifetime of an isolated atom, T_1. The reduction in the radiative lifetime is a consequence of the complete phase correlation of the individual emitters, and for this to be so it is necessary that the distance between them be comparable to the wavelength of the radiation.

The inhomogeneous broadening of the transition $\Delta\omega_D \sim 1/T_2^*$, however, leads to complete dephasing of the ensemble over a time T_2^* in the absence of the field. The phase shift for particles separated from the center of the inhomogeneous line by Δ over time τ_p is $\Delta\tau_p$ and, therefore, is different for the particles at different locations in the inhomogeneous line profile. This type of decay of a superradiant state is not due to relaxation processes and is therefore reversible in principle.

A (purely speculative) reversal of time in this system at time τ_p would evidently return the ensemble to its initial coherent state at time $2\tau_p$. An equivalent of time reversal is to have a π pulse, which rotates all vectors $\mathbf{R}_\Delta$ by an angle π, act on the system at time τ_p. The appearance at time $2\tau_p$ of a pulse of radiation due to reestablishment of a superradiant state is known as the photon echo effect.

Particle collisions, which break up the phase of the emitters, lead to irreversible dephasing of the system over a time T_2, so the amplitude of the photon echo signal decays exponentially as $\mathcal{E}_{echo} \sim \exp(-2\tau_p/T_2)$. We use just this fact to measure T_2.

The photon echo is also interesting because if we change the angle between the polarization directions of the exciting pulses then the polarization direction of the photon echo pulse also changes, in a way determined by the type and number of the transition forming the photon echo. This may be used to identify the type of transition.

The photon echo effect is also interesting because it allows us to study an interesting property of radiating systems (that is, cooperative effects) and to observe superradiant processes.

The observation and study of the photon echo in gaseous BCl_3 and SF_6 allowed us to measure the times T_2 for these gases and, therefore, to measure the homogeneous widths of the absorbing lines.

By studying the polarization dependence of the photon echo it was possible to identify the type of transition responsible for the echo in SF_6 at a frequency of 947.73 cm^{-1}, i.e., the frequency of the P16 line of a CO_2 laser, as a transition in the Q-branch with a high value of angular momentum J.

During a study of the concentration dependence of the photon echo signal the quadratic dependence of the photon echo on the concentration of gas particles was established. A range of pressures was isolated in which the decay of the superradiant state immediately after passage of the first exciting pulse becomes important and the radiative lifetime of the superradiant state was found to be $T_s \approx 4$ nsec · Torr (in our experimental geometry).

The third chapter of this paper deals with a study of the nutation effect in BCl_3 and SF_6. The nutation effect, as the spin echo effect, was discovered in NMR work by Torrey [14]. The name of the effect follows from the widely used NMR-classical analogy between an ensemble of spins in a constant magnetic field with a small, varying applied, magnetic field and a rotating top whose axis precesses due to a constant perturbation (an analog of the transition frequency) and slowly rotates (nutates) due to a harmonic perturbation in resonance with the precession frequency.

The possibility of an optical analog of nutation was predicted by Tang [15] who discussed the possibility of observing this effect in a CO_2 amplifying medium as a CO_2 laser pulse propagates in it. Nutation is easily understood by using the vector model employed to explain the photon echo (Fig. 1). We consider a perturbation in the form of a step in the electric field of a light wave at a frequency resonant with the transition frequency; that is, we shall not limit the duration of the perturbation. In this case the vector **R** slowly (compared to the field frequency) rotates at frequency $\Omega = \mu \mathcal{E}/\hbar$, and its transverse component, the polarization of the medium, varies harmonically as $\sin \Omega t$ with the same frequency Ω, proportional to the amplitude of the perturbing field and to the matrix element of the dipole moment of the transition. The polarization created by the field reradiates, and the resulting field which passes through the medium is modulated at frequency Ω. This effect is known as nutation and the frequency Ω is known as the nutation frequency.

In experimental study of nutation, one is faced with the purely technical difficulty of obtaining a perturbing field in the form of a step in the electric field of the light wave. Hocker and Tang [7], who first observed the optical analog of nutation (in a CO_2 laser–SF_6 system), used a CO_2 laser pulse (the P18 line) as a perturbing field and found the period of the observed nutations to be of the same order of magnitude as the risetime of the perturbing pulse. Apparently, as a result of just this, the dipole moment of the absorbing transition measured there [7] for the gas at the frequency of the P18 CO_2 laser line was $3 \cdot 10^{-20}$ cgse (the forbidden transitions of the 100-001 bands in CO_2 have just such a tiny dipole moment).

We have proposed a new method of observing nutation which is based on the fact that nutation occurs at a constant CO_2 laser power level immediately after irradiation of the medium with a short CO_2 laser pulse of the same frequency, resonant with the transition frequency but more intense. Then the pulse plays the role of a short (compared to the nutation period) perturbation pulse and the arrangement for observing nutation approaches optimal, which is when the perturbation is a step in the electric field of the light wave.

We have made a theoretical analysis of nutation including degeneracy and broadening inhomogeneity of the resonant transition with $\Delta\omega_D \gg \Omega$ for the above-described technique for observing nutation. It is shown that damping of nutation in BCl_3 and SF_6 has three causes: relaxation with time T_2, and degeneracy and inhomogeneous broadening of the transition responsible for producing the nutation. In addition, the nature of the nutation damping varies depending on the type and number of the transition. Damping is maximal for a Q transition with a large number.

The experimental study of nutation in BCl_3 and SF_6 was made with an apparatus consisting of two lasers which can be tuned over the CO_2 emission lines. One of the lasers produces millisecond pulses and plays the role of a constant perturbation and the other produces short pulses of length 100-200 nsec. The sample gas is put in a 3-m-long cell. The radiation is detected by a specially calibrated Cd–Hg–Te detector with a sensitivity of $0.4\ V \cdot W^{-1} \cdot cm^2$.

The dependence of the nutation frequency on the magnitude of the perturbing field was studied for those transitions in SF_6 which are resonant with the P16, P18, and P20 transitions of a CO_2 laser, and in BCl_3, with the P12 line frequency. The nature of the nutation damping was studied and the dipole moments of these transitions were measured. Proceeding from these values of the dipole moments and the absorption coefficients (also measured at these frequencies in BCl_3 and SF_6), we have evaluated the absorption cross sections and relative number of particles participating in the absorption.

A detailed description of the experimental results on self-induced transparency, photon echo, and nutation is given in Chapters I, II, and III.

CHAPTER I

SELF-INDUCED TRANSPARENCY IN GASEOUS BORON TRICHLORIDE

This chapter is devoted to a study of self-induced transparency in gaseous BCl_3 on the vibrational–rotational transition resonant with the P12 line of a CO_2 laser. A study of this effect makes it possible to estimate the value of the dipole moment and the homogeneous width of the absorbing transition [16]. Although these parameters of the gas may be measured reliably using the photon echo and optical nutation effects, self-induced transparency, as the simplest effect from an experimental standpoint, makes it possible to isolate the duration and intensity ranges for pulses which are π pulses for the given transition. In addition, by changing the intensity of such a pulse it is possible to form a pulse with arbitrary θ and, in particular, a $\pi/2$ pulse, which is needed to observe the photon echo.

Thus, self-induced transparency is the first and simplest, as well as necessary, step in studying coherent interaction effects.

1. A Theoretical Description of the Effect

Self-induced absorption occurs when a pulse of monochromatic radiation propagates through a resonant absorbing medium and the pulse length is shorter than the relaxation time of the polarization induced by the field of the pulse. Then the presence of a time-dependent polarization in the medium is the dominant factor, and a description of the system solely in terms of a difference in populations with the aid of rate equations is inadequate.

A semiclassical approximation is used to discuss self-induced transparency theoretically when the medium, in our case a resonant absorbing gas, is a two-level system described quantum mechanically by a density matrix and the perturbing field of the pulse propagating in the medium is described classically by the wave equation.

The complete Hamiltonian of the system in the absence of relaxation and degeneracy has the form

$$\hat{H} = \hat{H}_0 + \hat{\mu}\mathbf{E}, \tag{2}$$

where $\hat{H}_0 = \frac{\hbar\omega}{2}\begin{Vmatrix} 1 & 0 \\ 0 & -1 \end{Vmatrix}$ is the unperturbed Hamiltonian of a two-level system with particle density n and transition frequency ω_0. The perturbation $\hat{\mu}\mathbf{E}$ is caused by transitions due to the field from a plane-wave light pulse as it passes through the medium in the **z** direction polarized along the **x** axis:

$$\mathbf{E}(z, t) = \mathbf{x}\mathcal{E}(z, t)\cos(\omega t - kz), \tag{2a}$$

where the envelope $\mathcal{E}(z, t)$ varies slowly compared to the field frequency ω, i.e.,

$$\left|\frac{\partial\mathcal{E}}{\partial t}\right| \ll \omega|\mathcal{E}|; \quad \frac{\partial\mathcal{E}}{\partial z} \ll \left|\frac{\mathcal{E}}{\lambda}\right|. \tag{3}$$

Here $\hat{\mu}$ is the dipole moment operator with nonzero diagonal matrix elements, $\mu = \langle 1|\hat{\mu}|2\rangle$.

The equation for the density matrix is

$$i\hbar\dot{\rho}(t) = [\hat{H}, \rho(t)] \tag{4}$$

and may be rewritten in expanded form as

$$\begin{gathered} \dot{\rho}_{12} + i\omega_0\rho_{12} = -\frac{i}{\hbar}(\rho_{11} - \rho_{22})\mu E, \\ \text{complex conjugate for } \rho_{21}, \\ \dot{\rho}_{11} - \dot{\rho}_{22} = \frac{2i}{\hbar}(\rho_{12} - \rho_{21})\mu E. \end{gathered} \tag{5}$$

The diagonal elements are proportional to the populations of the upper and lower states,

$$\begin{aligned} & n_1 = \rho_{11} n, \\ & \rho_{11} + \rho_{22} = 1, \\ & n_2 = \rho_{22} n, \quad n_1 - n_2 = n(\rho_{11} - \rho_{22}). \end{aligned} \tag{6}$$

The nondiagonal elements determine the macroscopic polarization of the medium,

$$\mathbf{P}(z, t) = n\,\mathrm{Tr}[\hat{\mu}\rho] = n\mu[\rho_{12} - \rho_{21}]\,\mathbf{x}. \tag{7}$$

We shall represent the polarization **P** in the form of a slowly varying amplitude,

$$\mathbf{P}(z, t) = U(z, t)\cos(\omega t - kz) + V(z, t)\sin(\omega t - kz), \tag{8}$$

where U and V are the in-phase (or dispersion) and out-of-phase (or absorption) components, respectively, of the polarization. We also introduce the function $W(z, t) = {}^1\!/_2 n\hbar\omega(\rho_{11} - \rho_{22})$, which is a measure of the energy stored in the two-level system. In terms of these physical parameters Eqs. (5) take the form

$$\begin{aligned} & \dot{U} = V\Delta - \frac{U}{T_2}, \\ & \dot{V} = -U\Delta - \frac{2\mu^2}{\hbar\omega}\mathcal{E}W - \frac{V}{T_2}, \\ & \dot{W} = \frac{V\mathcal{E}\omega}{2} - \frac{W - W_0}{T_1}, \end{aligned} \tag{9}$$

where Δ is the difference between the frequency of the perturbing field of the pulse and the transition pulse, $\omega - \omega_0$. Equations (9) are an electrical analog of the Bloch equations [1]. As in the Bloch equations, here the relaxation terms are introduced phenomenologically assuming that the polarization components U and V are damped with a characteristic time T_2 and that the function W, defined by the population difference, approaches an equilibrium value $W_0 = -n\hbar\omega/2$ over a characteristic time T_1 in the absence of a field.

In a gaseous medium the relaxation mechanism is collisions of the gas molecules. The relaxation over time T_1 is produced by so-called inelastic collisions, or collisions which change the energy of a molecule, and relaxation over time T_2 is produced by collisions which shift the phase of the radiating molecules without changing their energy, or elastic collisions, as well as by inelastic collisions. Thus, the time T_2 is always less than or equal to T_1.

For an exact resonance ($\Delta = 0$) and in the absence of relaxation (T_1, $T_2 \to \infty$) the solution of these equations has the form

$$V(z, t) = n\mu \sin\frac{\mu}{\hbar}\varphi(z, t), \tag{10}$$

$$W(z, t) = -\frac{n\hbar\omega}{2}\cos\frac{\mu}{\hbar}\varphi(z, t), \tag{11}$$

where $\varphi(z, t) = \int\limits_{-\infty}^{t} \mathcal{E}(z, t')\,dt'$.

An important difference between problems on the coherent interaction of a light field with matter and nuclear magnetic resonance (NMR) problems is the need to take the propagation of the radiation into account. Then the field of a light wave [Eq. (2a)] pulse propagating in the medium along the z axis obeys the equation

$$\frac{\partial^2 E(z,t)}{\partial z^2}-\frac{1}{c^2}\frac{\partial^2 E(z,t)}{\partial t^2}=\frac{4\pi}{c^2}\frac{\partial^2 P(z,t)}{\partial t^2}. \tag{12}$$

Including the condition (3) that the envelope of the pulse varies slowly, the wave equation (12) takes the form

$$\frac{\partial \mathcal{E}(z,t)}{\partial z}+\frac{\partial \mathcal{E}(z,t)}{\partial t}=-\frac{2\pi\omega}{c}V(z,t). \tag{13}$$

The joint solution of Eqs. (13) and (9) for self-induced transparency obtained by McCall and Hahn [4] reduces to the area theorem

$$\frac{d\theta}{dz}=-\frac{\alpha}{2}\sin\theta, \tag{14}$$

where the parameter $\theta(z)=\frac{\mu}{\hbar}\int_{-\infty}^{\infty}\mathcal{E}(z,t)\,dt$ is proportional to the area under the pulse $A=\lim_{t\to\infty}\varphi(t)=\int_{-\infty}^{\infty}\mathcal{E}(z,t')\,dt'$ and to the matrix element of the dipole moment of the transition μ, while the constant α is the linear absorption coefficient,

$$\alpha=\frac{4\pi n\omega_0 T_2\mu^2}{\hbar c}. \tag{15}$$

Equation (14) is obtained by substituting Eq. (10) into (13) and integrating it over time. Also on multiplying Eq. (13) by $\frac{c}{4\pi}\mathcal{E}(z,t)$ and integrating with respect to time from $-\infty$ to ∞ we can also use Eqs. (13) and (10) to obtain an equation describing the energy loss from the pulse,

$$\frac{dS}{dz}=\frac{n\hbar\omega}{2}(\cos\theta-1), \tag{16}$$

where $S(z)=\frac{c}{8\pi}\int_{-\infty}^{\infty}\mathcal{E}^2(z,t')\,dt'$ is the energy of the pulse.

The solution of Eq. (14) is illustrated in Fig. 2 and has the form

$$\theta(z)=2\tan^{-1}\left[\tan\frac{\theta(z=0)}{2}\exp\left(-\frac{1}{2}\alpha z\right)\right]. \tag{17}$$

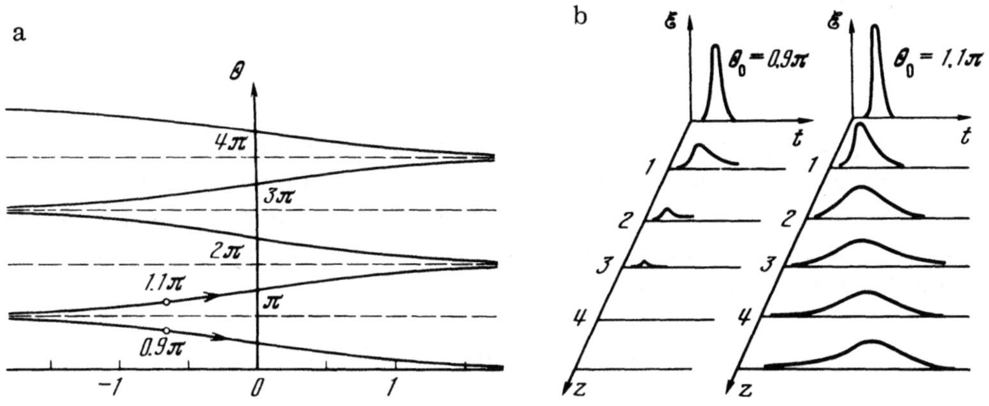

Fig. 2. An illustration of the solution, Eq. (17), of Eq. (14) for $\theta(z)$ for various values of θ_0 of the incident pulse (a) and the variation in the shape of the incident pulse with θ_0 greater than ($\theta_0 = 1.1\pi$) and less than ($\theta_0 = 0.9\pi$) the critical value $\theta_0 = \pi$ as it propagates in the medium (b).

Thus, an incident pulse with a value of θ (z = 0) somewhat greater than critical, θ (z = 0) = π, will broaden. Then the value of θ increases to 2π at distance $z \gg \alpha^{-1}$ and the pulse assumes the stable shape

$$\mathcal{E}(z, t) = \frac{2\hbar}{\mu\tau_{\text{pulse}}} \operatorname{sech}\left[\frac{1}{\tau_{\text{pulse}}}\left(t - \frac{z}{v}\right)\right]. \tag{18}$$

Furthermore the pulse propagates at a constant velocity $v < c$, determined by the parameters of the two-level system without energy loss. Then that part of the energy which leaves the first half of the pulse due to absorption, and, therefore, to excitation of the system, is compensated by the same amount of energy radiated by the system in the second half of the pulse. The pulse is delayed by a delay time, $\tau_{\text{delay}} = {}^1/_2 \alpha z \tau_{\text{pulse}}$ [4]. In its pure form the self-induced transparency effect consists of just this.

The delay in a π pulse, that is, one with θ (z = 0) = π, is much greater. The delay may easily be estimated by assuming that θ near π changes slowly and the energy of the pulse decays exponentially; then, $\tau_{\text{delay}} \simeq {}^1/_2 \tau_{\text{pulse}} \exp(\alpha z/2)$.

On the right side of Eq. (14), which is valid for a nondegenerate system, there is an oscillatory function. Thus, pulses with θ equal to 3π, 5π,..., will behave similarly to a π pulse. However, if degeneracy [10], which is always involved in the vibrational–rotational transitions of molecules, is included, the situation is considerably modified.

The vibrational–rotational transitions of molecular gases are degenerate in the projection m of the total angular momentum J in a given direction in space, here the direction of the field. The dipole moment of the transition between degenerate states depends on the number m and for transitions due to a linearly polarized field the selection rule $\Delta m = 0$ holds. The dependence of μ on m and J for (0Jm – 1J'm) transitions where $J' = J \pm 1$, J (the P, R, and Q branches, respectively, of the vibrational–rotational transitions with a change in the vibrational quantum number by unity) in a symmetric-top molecule has a universal character and is given by the formulas

$$\mu_m = (m/J)\,\mu_0, \tag{19}$$

$$\mu_m = (\sqrt{J^2 - m^2}/J)\,\mu_0. \tag{20}$$

Including degeneracy and assuming that all degenerate states are equally probable and, therefore, equally populated with a density $n/(2J + 1)$, the area theorem may be written in the form

$$\frac{d\theta}{dz} = -\frac{\alpha}{2(2J+1)} \sum_{m=-J}^{J} \frac{\mu_m}{\mu_0} \sin\left(\frac{\mu_m}{\mu_0}\theta(z)\right). \tag{21}$$

The right side of Eq. (21) is proportional to the polarization per unit volume of a degenerate medium and is determined by the total contribution to the polarization from the various degenerate states $m = 0, \pm 1, \pm 2, \ldots \pm J$; μ_0 is the maximum magnetic moment; μ_m is the matrix element of the m-th degenerate state; and the parameter θ and constant α are defined in terms of the maximum matrix element μ_0 by

$$\theta(z) = \frac{\mu_0}{\hbar} \int_{-\infty}^{\infty} \mathcal{E}(z, t)\,dt; \qquad \alpha = \frac{4\pi n \omega \mu_0^2 T_2}{\hbar c}.$$

Similarly, taking the degeneracy into account, Eq. (16) is rewritten in the form

$$\frac{dS}{dz} = \frac{n\hbar\omega}{2(2J+1)} \sum_{m=-J}^{J} \left(\cos\frac{\mu_m}{\mu_0}\theta - 1\right), \tag{22}$$

where the right side is determined by the difference in the populations of the degenerate system due to a pulse with parameter θ.

Upon summing in the case of a transition in the Q branch with large J, when the distribution of the dipole moment over the degenerate states may be regarded as continuous, Eqs. (21) and (22) take the form

$$\frac{d\theta}{dz} = -\frac{\alpha}{2}\left(\frac{\sin\theta}{\theta^2} - \frac{\cos\theta}{\theta}\right), \tag{23}$$

$$\frac{dS}{dz} = \frac{n\hbar\omega}{2}\left(\frac{\sin\theta}{\theta} - 1\right), \tag{24}$$

as illustrated in Fig. 3. For comparison the dashed curve shows the case of a nondegenerate system.

Therefore, the existence of degeneracies leads to rapid damping of the functions on the right sides of Eqs. (23) and (24) and, therefore, to damping of the oscillations in the polarization and in the population difference in the degnerate system as the pulse parameter θ increases. Analysis of the curves of Fig. 3 also shows that in a degenerate system a pulse cannot propagate without losing energy even in the absence of relaxation (dS/dz always negative). Nevertheless, for pulses with $\theta \approx \pi$ and for a medium with optical thickness equal to several absorption lengths (that is, until the losses due to degeneracy become too large) a degenerate system behaves similarly to a nondegenerate one.

The significant differences in the behavior of a degenerate system appear for pulses with $\theta \gg \pi$. After such a pulse passes, the macroscopic polarization of the medium approaches zero. The system as a whole ceases to reradiate due to the interference of individual degenerate states. The population difference approaches zero and the pulse energy loss approaches the constant value $dS/dz = -n\hbar\omega/2$. This behavior of a degenerate system as the intensity of the incident pulse is increased is called "effective saturation" in the terminology of Rhodes [10]. In this case saturation is understood to mean that a further increase in the incident intensity does not change the macroscopic state of a degenerate system while the absorption coefficient of the sample continues to decrease monotonically.

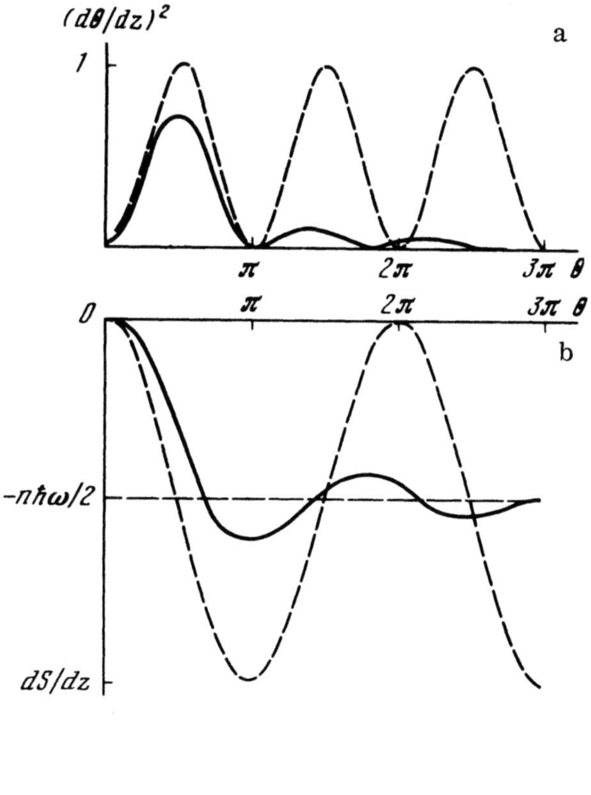

Fig. 3. Curves of $(d\theta/dz)^2$, which is proportional to the polarization per unit volume of the medium (a), and dS/dz, which describes the energy loss from the pulse (b), versus the pulse parameter θ. The continuous curves are for a degenerate system (transition in the Q branch with large $J \gg 1$) and the dashed curves are for a nondegenerate system.

Thus, by studying the propagation of pulses of coherent radiation in degenerate systems near the π-region, where degenerate and nondegenerate systems behave in a similar fashion, it is possible to experimentally evaluate the dipole moment of an absorbing transition by the expression $\mu \mathcal{E} \tau_{pulse} / \hbar \approx \pi$. In addition, when the inequality $\tau_{pulse} \ll T_2$ is artificially violated it is possible to determine the relaxation time T_2 from the way the shape of the pulse changes as it propagates in the medium.

2. The Experimental Apparatus

A diagram of the experimental apparatus for studying self-induced transparency in BCl_3 is shown in Fig. 4.

The source of pulses of coherent radiation is a CO_2 laser L working in a double Q-switched mode [17]. Q-switching is done using a rotating mirror R with frequency of 200-400 Hz synchronized with supply pulses of duration 1 msec and repetition rate 25-50 Hz.

Use of supplementary Q-swithching with the supply greatly increases the gain of the laser medium [18], so it is possible to work with short lasers (of order 1 m) with low-Q-selections in the cavity to ensure high stability of the output amplitude. In addition, this type of laser operates practically without pumpthrough of the mixture and does not require water cooling, so the experiments are much easier to perform.

The simplest design for a laser of this type and the one used is a glass tube of diameter 15-25 mm and length 1-2 m with sealed-in electrodes and stopcocks at the ends for feeding and pumping out the mixture. The tube is vacuum sealed with Brewster-angle NaCl windows, so the laser output is linearly polarized. A constant voltage of 10-20 kV is applied to the anode and a modulated tube is placed in the cathode circuit to produce the pulsed discharge. A premixed combination of the gases CO_2-N_2-He in the ratio 1 : 1 : 3 is pumped through the discharge tube. The gas pressure usually does not exceed 10 Torr. The flow rate of the mixture is very low so the laser can operate without pumpthrough for 10-15 min; however, then the intensity of the output gradually decreases due to dissociation of CO_2 molecules in the discharge [19].

The laser cavity is made up of a plane-parallel germanium plate with a dielectric coating, a rotating mirror R, and a diffraction grating G (100 lines/mm). The reflectivity of the Ge output mirror is 70% due to a multilayer dielectric coating on one of its faces. The other face of

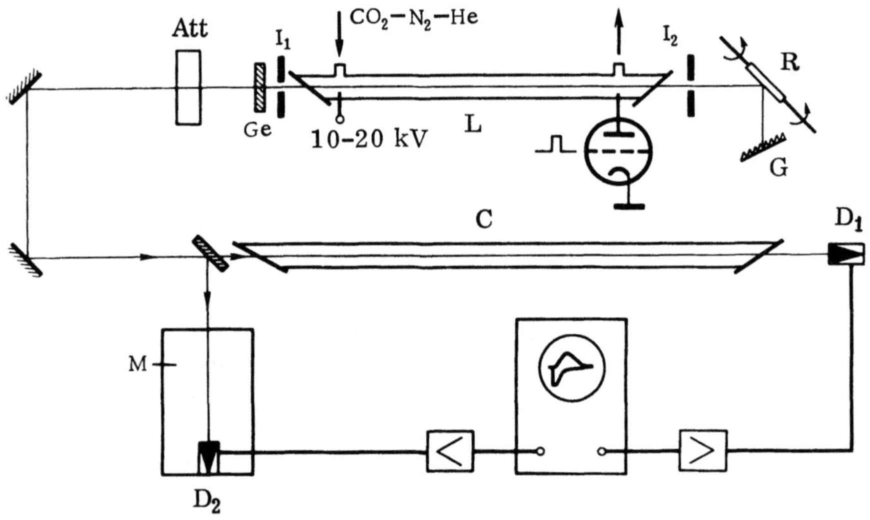

Fig. 4. A diagram of the experimental apparatus for studying self-induced transparency in BCl_3.

the output mirror is coated since otherwise the output mirror would act as an additional selective element in the cavity with a transmission coefficient determined by the mirror thickness. This would lead to a substantial instability in the output intensity even when the temperature of the surrounding air and, therefore, the thickness of the mirror underwent negligible oscillations.

The diffraction grating is used to tune the output over the laser lines. The use of double Q-switching greatly increases the tuning range of the laser, so it is possible to work the weak P8, P10, and P12 CO_2 laser lines where the test gas has the highest absorption coefficient.

Two iris diaphragms, I_1 and I_2, are placed inside the cavity to ensure laser operation in the fundamental transverse (TEM_{00}) mode and make it possible to change the length of the output pulses.

Among the shortcomings of this laser we must include a frequency instability within the homogeneous broadening limits of the CO_2 laser line (50 MHz at a pressure of 10 Torr) due to an instability in the length of the cavity, one of whose windows is the rotating mirror. This frequency instability is important when observing self-induced trasparency in BCl_3 and, especially, SF_6, since these gases have a very dense spectrum of absorption lines and the frequency instability of the laser leads to an instability in the amplitude of the signal passing through the cell with the absorbing gas. The mechanical instability in the frequency was substantially reduced by using a massive gyroscope as a rotating mirror and by smoothly tuning the frequency with a piezoelectric crystal mounted on one of the cavity mirrors.

This laser yields pulses of length 100-500 nsec with intensities of several kW/cm^2 at many CO_2 laser lines.

The laser pulses were deflected with mirrors into a cell C of length 3 m and diameter 40 mm containing the test gas. The cell was vacuum sealed with NaCl windows mounted at the Brewster angle and equipped with fittings for feeding in and pumping out the test gas. The cell was pumped down to the required pressure with a TsVL-100 diffusion pump and the cell was then closed. Because of the good vacuum sealing of the cell the gas pressure is maintained over several hours. The length of the cell was chosen so it would cover several absorption lengths α^{-1} of the test gas. The pressure in the cell is limited above by the fundamental requirement that the interaction be coherent; for pulses of duration 100 nsec, it does not usually exceed 0.1 Torr. The pressure was measured with a specially calibrated manometer tube (type PMT-2).

The laser output intensity was varied either with a step attenuator, Att, made of thin-film polyethylene with an absorption coefficient of $\alpha = 30\ cm^{-1}$ at a wavelength of 10 μ, or with two crossed polarizers (diffraction gratings with 1200 lines/mm mounted on thin transparent films).

An attenuator of the first kind has a practically unlimited range of intensity variation; however, it bends and scatters the passing beam somewhat. The second type of attenuator avoids this shortcoming and is very convenient to use, but its attenuation is limited to 30 dB.

To monitor the frequency and intensity of the laser imput to the cell, a portion of the beam was deflected into an IKM-21 monochromator (M). The monochromator was equipped with a 100 lines/mm diffraction grating to increase its resolution; this makes it possible to reliably identify the CO_2 laser lines.

The radiation is detected by two IR detectors D_1 and D_2. Detector D_2 records the amplitude and shape of the input pulse, monitors the frequency of the radiation, and triggers the sweep on the S1-42 oscilloscope. Detector D_1 records the amplitude and shape of the pulse after it passes through the test gas in the cell.

Initially Ge–Zn–Sb detectors with a bandwidth of 30 MHz and relatively low sensitivity ($0.01\ V \cdot W^{-1} \cdot cm^2$) were used in the experiments. Then it was necessary to focus the radiation onto the detector surface with a short-focal-length NaCl lens in order to increase the signal from the detector. It is highly undesirable to focus the radiation in experiments on intensity threshold effects, including self-induced transparency, since the output of the laser is inhomogeneous over the perpendicular cross section. In later experiments it was possible to avoid focusing by using more sensitive Cd–Hg–Te detectors.

3. Experimental Results

The apparatus described in Section 2 was used to observe self-induced transparency in BCl_3. The frequencies of the vibrational–rotational absorbing transitions of this gas coincide, to within the Doppler width of the line, with the frequencies of a number of CO_2 laser lines. The effect was investigated at a frequency of 951.16 cm^{-1}, which corresponds to the P12 line of the 001-100 vibrational–rotational band of the CO_2 laser. At this frequency BCl_3 has a maximum absorption coefficient of $\alpha = 0.13\ cm^{-1} \cdot Torr^{-1}$.

The intensity, shape, and delay of the output pulse as functions of the input intensity were studied experimentally for various pulse lengths with gas pressures in the cell of from 1 to 100 mTorr (Fig. 5). A bend can easily be seen in the curves of Fig. 5 at a location I_b which depends on the duration of the input pulse. The quantity I_b shifts toward larger input intensities as the duration of the pulse is reduced (Fig. 5a) and is independent of the gas pressure (Fig. 5b). That the location of the bend I_b depends on the pulse length is an indication that the threshold for the effect is determined both by the intensity and by the duration of the input pulse; that is, it is determined by the area under the pulse.

For radiant intensities below threshold $I < I_b$, the pulse is absorbed linearly according to the Bouguer–Beer law and its shape is not distorted. For intensities much greater than threshold ($I > I_b$) the medium practically becomes transparent and $I_{out} \rightarrow I_{in}$. However, this process is accompanied by the loss of some energy from the leading edge of the pulse needed to excite the medium.

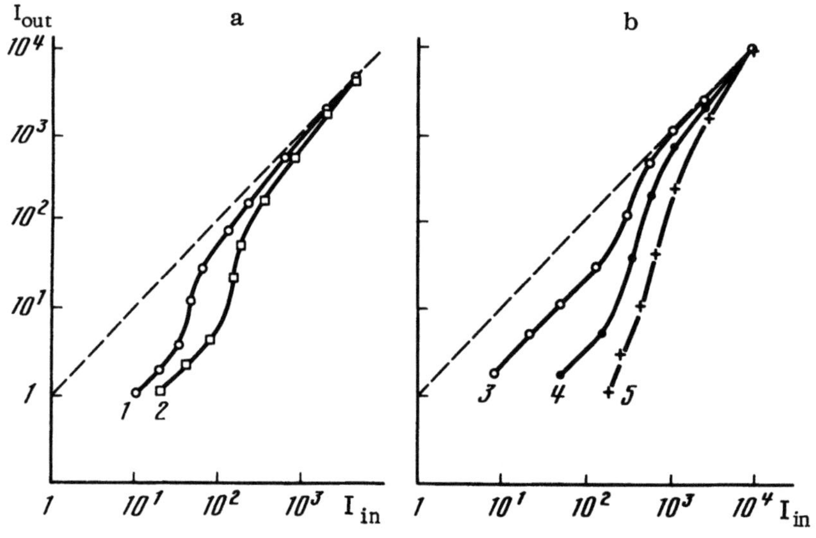

Fig. 5. Dependence of the output intensity I_{out} on the input intensity I_{in} (in relative units). a) For various pulse lengths: 1) τ_{pulse} = 360 nsec; 2) 80 nsec; b) For various gas pressures in the cell; 3) P = 40 mTorr; 4) 75 mTorr; 5) 120 mTorr.

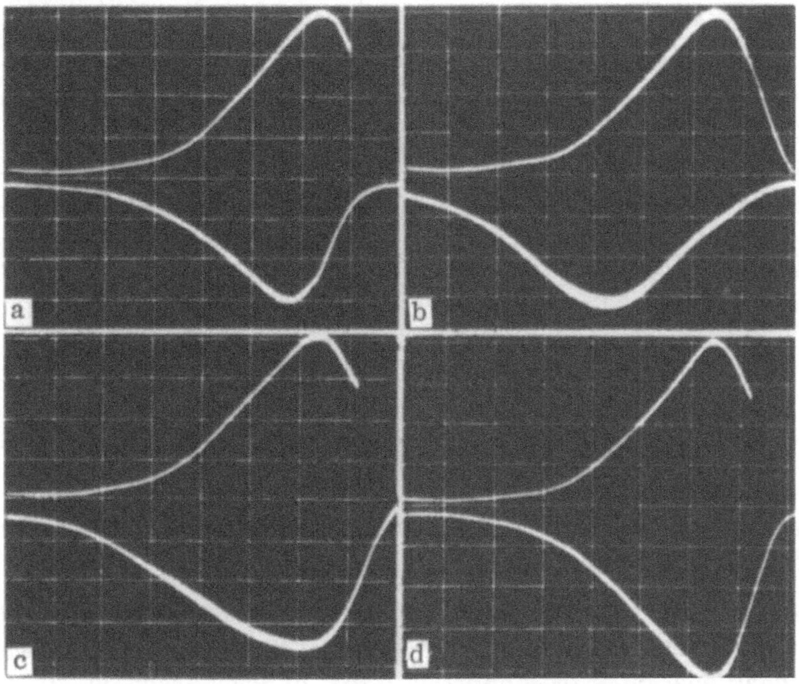

Fig. 6. Photographs of a series of output pulses with increased input pulse intensities. Pulse (c) corresponds to a π pulse. The lower traces show the input pulses. The time scale is 50 nsec.

The dependence of the shape and delay of the output pulse on the input intensity is illustrated in Fig. 6. For pulse intensities much lower than threshold the output pulse shape is not distorted and there is no delay (Fig. 6a). The pulse shape changes considerably at intensities in the neighborhood of $I = I_b$. The pulse broadens, becomes symmetric, and its delay is maximal (Fig. 6c). As the input pulse intensity is further increased, the output pulse again reproduces the input pulse shape and the delay disappears (Fig. 6d). Then, as already noted above, the pulse loses part of its energy from the leading edge, an energy balanced by the population difference in the system. The trailing edge of the pulse is unchanged. According to Section 1, the amount of energy lost per unit volume of the medium is $n\hbar\omega/2$, and, therefore, is determined by the gas pressure. This process is illustrated in Fig. 7, which was made by superimposing two oscilloscope traces of the output pulse with intensity $I \gg I_b$ obtained when

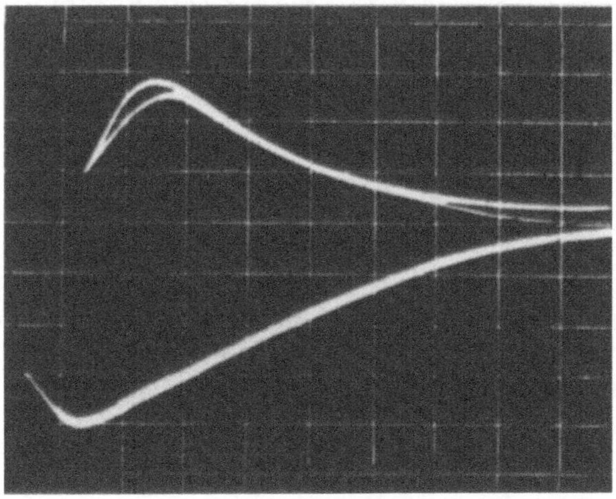

Fig. 7. A superposition of two oscilloscope traces of an output pulse with $\theta \gg \pi$ when the cell is fully evacuated and when it has a gas pressure of 100 mTorr. The input pulses shown on the lower trace coincide. The time scale is 50 nsec.

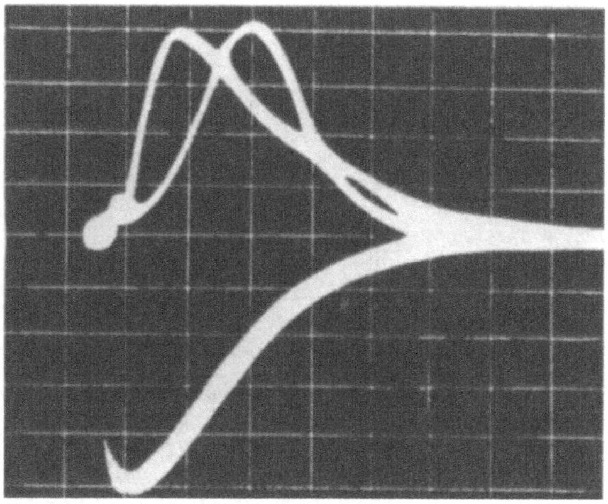

Fig. 8. A superposition of two oscilloscope traces of an output pulse with $\theta \approx \pi$ when the cell is completely pumped out and when it has a BCl_3 pressure of 100 mTorr. The input pulses shown in the lower trace coincide. The time scale is 100 nsec.

the cell is completely pumped out and when it has a gas pressure of 100 mTorr. The input pulses shown on the lower trace coincide.

This pulse behavior at intensities $I \gg I_b$, the absence of additional bends in the curves of Fig. 5, and the characteristic distortion in the shape of the output pulse in the threshold region are experimental confirmation of the important effect of degeneracy on self-induced transparency in BCl_3.

The sharp change in the absorption coefficient and the distortion and maximum delay in the output pulse near the bend (at $I = I_b$) suggest that this bend (in the curves of Fig. 5) corresponds to a value of the output pulse parameter of $\theta = \pi$. Calorimetric measurements of the absolute output intensity at the bending point and measurements of the pulse duration make it possible to estimate the value of the dipole moment of this transition in the test gas using the relation

$$\frac{\mu}{\hbar}\sqrt{\frac{8\pi}{c} I_b}\, \tau_{pulse} \approx \pi. \qquad (25)$$

Here the real degenerate system approaches a nondegenerate system and the (matrix element of the) dipole moment of the transition that is being evaluated for a nondegenerate system is a sort of effective dipole moment for the real system.

From these estimates we obtain the value $\mu = 5 \cdot 10^{-20}$ cgse for the vibrational–rotational transition in BCl_3 responsible for absorption of the P12 CO_2 laser line.

The deformation and delay in the output pulse near the bend in the curves (see Fig. 5) are characteristic of the case when the pulse duration is less than the relaxation time T_2.

As the pressure is increased, and, therefore, T_2 is decreased, there is no longer a distinct bend in the curves of Fig. 5. In addition, the way the shape of the output pulse changes becomes different. As before, the pulse is delayed due to absorption of its leading edge by the medium; however, since the reverse process of reradiation takes place only during a time T_2 while the molecules have not destroyed the polarization of the medium induced by the field of the pulse, the leading edge of the pulse is shortened. This process is illustrated in Fig. 8, which was obtained by superimposing two oscillosope traces of the output pulse with $\theta \approx \pi$ when the cell was pumped out and when it had a BCl_3 pressure of 100 mTorr. From experiments of this type it was possible to estimate the time T_2 to be about 300 nsec at a BCl_3 pressure of 100 mTorr, or $T_2 \approx 30$ nsec · Torr.

Thus, an experimental study of the propagation of CO_2 laser pulses in gaseous boron trichloride has established the effect of degeneracy on self-induced transparency in BCl_3 and has isolated the ranges of pulse duration and intensity in which the behavior is similar to the behavior of a π pulse in a nondegenerate two-level system. This makes it possible to evaluate the effective dipole moment of the transition. In addition, the transverse relaxation time T_2 of this gas has been estimated from the way the shape of the output pulse changes when the inequality $\tau_{\text{pulse}} < T_2$ is violated.

CHAPTER II

THE PHOTON ECHO IN BCl_3 AND SF_6

This chapter is devoted to a study of the photon echo in gaseous BCl_3 and SF_6 when they are irradiated with pulses of monochromatic infrared radiation from a CO_2 laser separated by a time interval τ_p [20]. A necessary condition for observation of the photon echo is that the interval τ_p be small compared to the relaxation time T_2 of the polarization induced in the medium by the field of the exciting pulses. Thus, as in the case of self-induced transparency, the existence of a time-dependent polarization of the medium is the determining factor.

The photon echo effect is one of the most interesting coherent interactions since with it one can reliably measure the homogeneous widths of absorbing transitions of the gases being examined here, identify the types of these transitions [20], and study one of the most interesting properties of radiating systems (cooperative effects) as well as observe superradiant processes [21].

4. The Theory of the Effect

We assume that a pulse of plane, linearly polarized light of the form given in Eq. (2), with an envelope $\mathcal{E}(z, t)$ satisfying the condition $\theta(z=0) = \frac{\mu}{\hbar}\int\limits_{-\infty}^{\infty}\mathcal{E}(0, t)\,dt = \frac{\pi}{2}$ (where μ is the matrix element of the dipole moment of the transition), with duration $\tau_{\text{pulse}} \ll T_2$, and propagating along the x axis, is incident on the medium. This kind of pulse produces maximum polarization of the medium in the propagation region. A collection of particles driven by the $\pi/2$ pulse into the upper excited state is an ensemble of dipoles for which the remarks below are true:

1. This ensemble will include only those particles in the volume of the medium for which the laser frequency is resonant, that is, for laser light whose frequency falls within the interval $\omega_0 \pm \Delta$ due to the Doppler effect, where ω_0 is the center of the Doppler absorption line and the width Δ is defined in our case ($\tau_{\text{pulse}} \ll T_2^*$) as the width of the pulse spectrum $1/\tau_{\text{pulse}}$ if $\mu\mathcal{E} \ll 1/\tau_{\text{pulse}}$, and as the width of the Doppler line profile $1/T_2^*$ if $\mu\mathcal{E} \gg 1/\tau_{\text{pulse}}$. For a $\pi/2$ pulse, $\mu\mathcal{E} \sim 1/\tau_{\text{pulse}}$, hence $1/\tau_{\text{pulse}} < \Delta < 1/T_2^*$.

2. Over a time small compared to the transverse relaxation time T_2, the dephasing of the dipole oscillations due to their frequency distribution within the interval $\omega_0 \pm \Delta$ causes the ensemble to lose its coherence and, therefore, causes the disappearance of dipole radiation from the system almost immediately after the exciting pulse passes through the medium.

An important circumstance in this process is the fact that the decay of the coherent state of the dipole ensemble, which is due to this inhomogeneous broadening mechanism, has nothing to do with relaxation processes and is therefore reversible in principle. Time reversal in this system at some time τ_p (reckoning time from the first exciting pulse) evidently returns the ensemble to its initial coherent state at time $2\tau_p$. An equivalent of this "time reversal" is to act on the ensemble at time τ_p with a perturbation which transforms the polarization of the particles $\mathbf{P}_\omega$ into $-\mathbf{P}_\omega$. A perturbation which produces this reorientation of the polarization may be realized by sending a second exciting pulse into the medium at time τ_p with $\theta = \pi$. The appearance at time $2\tau_p$ of a radiation pulse caused by the restoration of the coherent state of the ensemble is known as the photon echo effect.

The theoretical description of the photon echo effect comes basically from the work of Hahn on the spin echo [2] and was developed by Kurnit et al. for ruby [9]. The more complicated case of photon echo in gases was examined in [11] and [12]. Gordon et al. [11] have examined the dependence of the intensity and polarization direction of the photon echo pulse on the angle between the polarization directions of the exciting pulses in a highly degenerate gaseous medium. This case corresponds most closely to the experimental case of the photon echo in BCl_3 and SF_6.

The ensemble of molecules is assumed to be a two-level system with degeneracy and is described by a density matrix ρ with elements $\langle vJKm|\rho|v'J'K'm'\rangle$. The set of quantum numbers v, J, and K characterizes the upper and lower levels which are degenerate in the projection m of the rotational angular momentum J in a given direction, for example, that of the field of the exciting pulses. The quantity v is the vibrational quantum number and K is the projction of the angular momentum J along the axis of symmetry of the molecule.

The equation of motion of the density matrix in the absence of relaxation (T_2, $T_1 \to \infty$) has the form

$$i\hbar\dot{\rho} = [\hat{H}_0 + \mathbf{E}\hat{\mu},\ \rho], \tag{26}$$

where $\hat{\mu}$ is the dipole moment operator with elements $\langle 1, m|\hat{\mu}|2, m'\rangle$. In solving Eq. (26) Gordon et al. [11] made two basic simplifying assumptions. The field of the exciting pulses was assumed to be the only external field seen by the molecules and to be constant over the length of the entire sample; that is, the secondary effect of the field on the polarization (caused by the polarization of the medium) was neglected. Thus, the results of the calculation are limited to the case of an optically thin medium where the intensity of the photo echo is very much less than that of the exciting pulses. Alekseev and Evseev [12] calculated the polarization dependence including the inverse process, but, unfortunately, only for small values of J. They showed that including the inverse process distorts the polarization properties of the photon echo.

The second assumption in [11] is to neglect the frequency difference Δ at the time the exciting pulses act; this is equivalent to the condition $\mu\mathscr{E}/\hbar \gg \Delta$. This assumption is certainly an idealization under our experimental conditions ($\mu\mathscr{E}/\hbar \approx \Delta$); nevertheless, this does not affect the results on the polarization dependence of the photon echo, although it surely distorts the dependence of the photon echo intensity on the intensity of the exciting pulses. That the assumption $\mu\mathscr{E}/\hbar \gg \Delta$ has little effect on the polarization properties of the photon echo is demonstrated, for example, by the agreement between [11] and [12] for small values of J since no limitation is placed on Δ at the time of application of the exciting pulses in [12].

Thus, solution of the photon echo problem reduces to solving Eq. (26) with these assumptions and to determining the polarization of the medium at time $2\tau_p$ (which is the polarization responsible for the appearance of the photon echo signal at that time) by following the standard procedure,

$$\mathbf{P} = \mathrm{Tr}[\hat{\mu}\rho(2\tau_p)]. \tag{27}$$

Here $\rho(2\tau_p)$ is the solution of Eq. (26) at time $2\tau_p$. Finally, the expression

$$\mathbf{P} = \mathrm{Re}\frac{i\exp[-i\Delta(t-2\tau_p)]}{2(2J+1)}\mathrm{Tr}[\hat{\mu}\sin\mathbf{A}_2\hat{\mu}\cdot\sin 2\mathbf{A}_1\hat{\mu}\cdot\sin\mathbf{A}_2\hat{\mu}] \tag{28}$$

is independent of the frequency difference Δ at $t = 2\tau_p$. Here the magnitudes of the vectors $\mathbf{A}_1$ and $\mathbf{A}_2$ are determined by the areas under the first and second pulses, respectively. The directions of these vectors are the same as the directions of the fields of the exciting pulses which themselves are rotated by some angle ψ about one another; i.e.,

$$\mathbf{A}_1 = \mathbf{x} \int_{\substack{\text{1st}\\ \text{pulse}}} \mathcal{E}(t)\,dt; \qquad \mathbf{A}_2 = (\mathbf{x}\cos\psi + \mathbf{y}\sin\psi) \int_{\substack{\text{2nd}\\ \text{pulse}}} \mathcal{E}(t)\,dt.$$

The intensity and polarization direction of the photon echo pulse as functions of the angle ψ are completely determined by the trace in the first part of Eq. (28),

$$Q \equiv \mathrm{Tr}[\hat{\mu}\sin \mathbf{A}_2\hat{\mu}\cdot\sin 2\mathbf{A}_1\hat{\mu}\cdot\sin \mathbf{A}_2\hat{\mu}]. \tag{29}$$

It is this expression which was analyzed in [11]. Figure 9 shows computed plots of the dependence of the polarization direction (a) and intensity (b) of the photon echo on the angle between the polarization directions on the exciting pulses, ψ, for various types of vibrational transitions (Q, P, and R branches) and various values of the angular momentum J.

From the curves of Fig. 9a it is clear that the polarization direction of the photon echo behaves differently depending on the type and number of the transition. For the P1 and Q1 transitions the photon echo polarization follows the polarization of the second pulse (that is, the angle between the polarization directions of the echo and the second pulse φ is zero), and the intensity of the echo pulse varies as $Q^2(\psi) = Q^2(0)\cos^2\psi$. However, for transitions with $J > 1$ the echo polarization is a more complicated function of the angle $Q^2(\psi) = Q^2(0)\cos^2\psi$. For Q-branch transitions ($\Delta J = 0$) the photon echo polarization lies within the angle between the polarization directions of the pulses ψ ($\varphi = 0$), and for the P- and R-branch transitions ($\Delta J = \pm 1$) it lies outside this angle ($\varphi > 0$).

We shall analyze the dependence of the intensity of the photon echo on the intensity of the exciting pulses, beginning with Eq. (29), which was obtained in [11], and assuming that $\psi = 0$

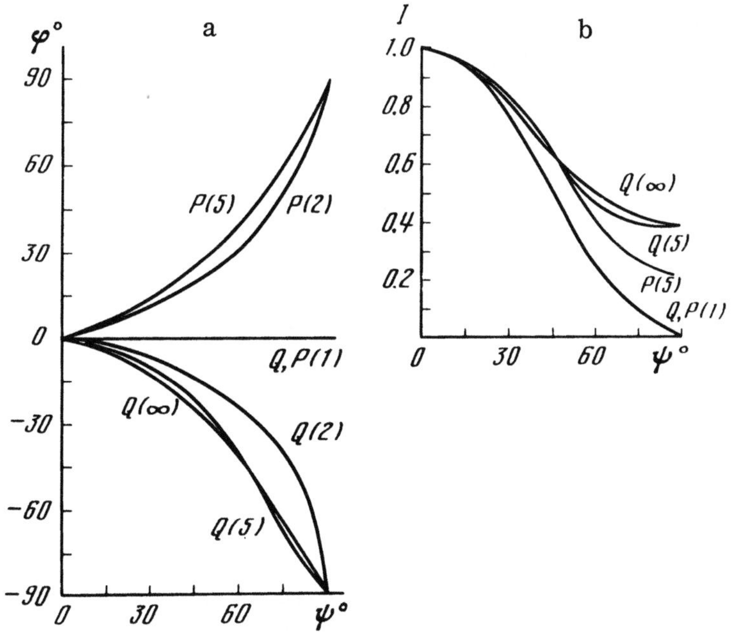

Fig. 9. The computed polarization dependence of the photon echo. a) Dependence of the angle φ (between the polarization directions of the echo and the second exciting pulse) on the angle between the polarization directions of the exciting pulses, ψ; b) dependence of the photon echo intensity on ψ for various types of transitions used to produce an echo.

and $A_2 = 2A_1$. We shall limit ourselves to the case when the transition producing the echo is a Q-branch transition with large J since this case is evidently closest to the actual one in BCl_3 and SF_6. As we did when taking degeneracy into account in self-induced transparency, we shall turn to a quasiclassical description assuming the dipole moment distribution over the degenerate states to be continuous with $\mu = \mu_0 \cos\beta$ ($\cos\beta = m/J$). Going from the summation in Eq. (29) to an integral over all possible orientations of the dipole moment, we obtain

$$Q = x\frac{\mu_0}{2}\int_{-1}^{1} \cos\beta \cdot \sin^3(2\theta_1 \cos\beta)\, d(\cos\beta). \tag{30}$$

Here the parameter of the exciting pulse is defined in terms of the maximum dipole moment μ_0: $\theta_1 = (\mu_0/\hbar)A_1$, $\theta_2 = (\mu_0/\hbar)A_2$.

The result of the integration,

$$|Q|^2 \sim \left[\frac{3}{8}\left(\frac{\sin 2\theta_1}{2\theta_1^2} - \frac{\sin 6\theta_1}{54\theta_1^2} - \frac{\cos 2\theta_1}{\theta_1} + \frac{\cos 6\theta_1}{9\theta_1}\right)\right]^2, \tag{31}$$

is plotted in Fig. 10, from which it is clear that the dependence of the photon echo intensity on the parameter θ_1 is an oscillating function which damps rapidly with increasing θ_1. This damping is inherent only in a degenerate system and is explained by the presence of a wide choice of values for the matrix elements of the dipole moment belonging to different degenerate states. These states behave the same way only for $\theta < \pi$. As θ is increased the macroscopic polarization of the medium, and, therefore, the photon echo itself, disappear due to interference of the contributions from the individual degenerate states to the polarization.

Up to now in our discussion of the photon echo we have neglected the effect of collisional, and, therefore, irreversible, dephasing of the molecular ensemble during the time interval $2\tau_p$ (from the time of the first exciting pulse until the photon echo signal appears), as is fully valid when $2\tau_p \ll T_2$. It is easy to obtain the dependence of the photon echo amplitude on τ_p including this mechanism (when $\tau_p \sim T_2$), keeping in mind the fact that the decay rate for a fully correlated ensemble is proportional to $1/T_2$. The amplitude of the echo pulse field as a function of the time separation between the exciting pulses in this case evidently has the form

$$\mathcal{E}_{echo} \sim \exp(-2\tau_p/T_2), \tag{32}$$

and its intensity obeys

$$I_{echo} \sim \exp(-4\tau_p/T_2). \tag{33}$$

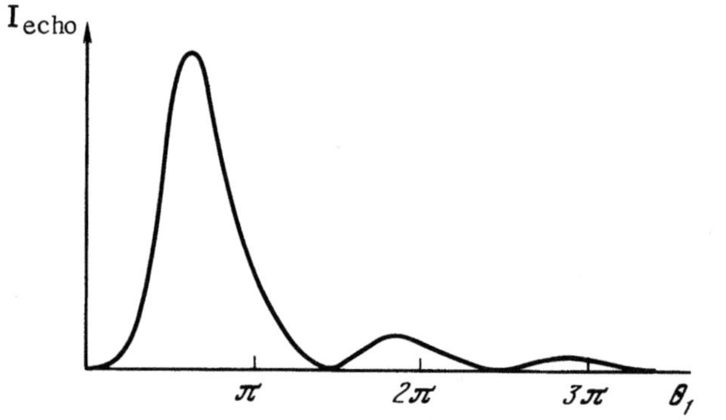

Fig. 10. Computed dependence of the photon echo intensity on the parameter θ_1 of the first exciting pulse.

Therefore, an experimental investigation of the photon echo effect in gases allows us to measure the collisional dephasing time T_2. In addition, from the form of the polarization dependence and from the dependence of the photon echo intensity on the intensities of the incident pulses it is possible to determine the type and number of the transition responsible for the photon echo.

5. Experimental Apparatus

The technique for observing the photon echo is to successively illuminate the medium with two light pulses separated by a time interval of length τ_p. To successfully study this effect the range of controllable variation of the interval τ_p must be much greater than the duration of the exciting pulses. When the photon echo was observed in ruby the same laser served as the source of both exciting pulses, the second pulse being formed with the aid of a variable optical delay line. The delays obtained in this manner were fully satisfactory since the lengths of the exciting pulses were several nanoseconds. Photon echo studies in a CO_2 laser–BCl_3 or SF_6 system with a minimum attainable Q-switched CO_2 laser pulse length of about 100 nsec require microsecond delays. The use of optical delay lines for these purposes is techically very difficult. Thus, in our case the exciting pulses were produced by different CO_2 lasers with synchronized emission times.

The experimental apparatus used to study the photon echo effect in BCl_3 and SF_6 is illustrated in Fig. 11. The apparatus consists of two pulsed CO_2 lasers, L and L', each of which has the same parameters as the laser described in Section 2 for observing self-induced transparency in these gases. The lasers have a single rotating mirror consisting of a plane parallel glass plate, gold coated on each side, in common. This allows synchronization of the emission time of the two lasers. Rotating the diffraction gratings about an axis perpendicular to the grooves makes it possible to smoothly vary the time interval between the output pulses over wide limits without retuning the laser frequency. The range of variation of the delay time depends on the aperture of the diffraction grating and the speed of rotation of the mirror; in our case it is 0-100 μsec. Rotating the gratings G and G' about (an axis parallel to) the grooves makes it possible to tune the lasers over the emission lines. The output pulses are attenuated by the attenuators (Att) and aimed into the cell C (length 3 m) containing the test gas by the bending mirrors 4, 5, and 6 and the beam splitter R. The gas pressure in the cell where the photon echo effect is observed is limited by the condition $\tau_p < T_2$ and usually lies in the range 10^{-3}-10^{-4} Torr. In some experiments a telescope T made of two NaCl lenses was mounted at

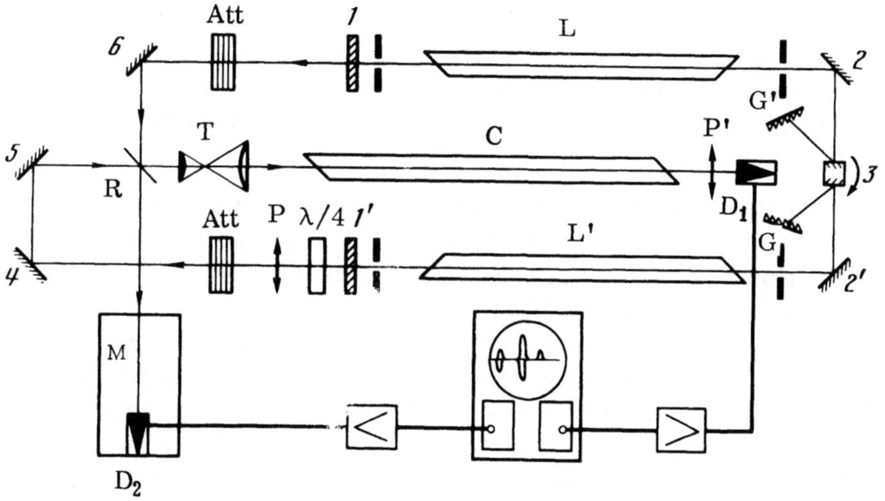

Fig. 11. A diagram of the experimental apparatus for studying the photon echo in BCl_3 and SF_6.

the entrance of the cell. The telescope magnifies the beam diameter and, therefore, the number of particles taking part in the effect, as well as compensating for diffractive divergence and losses due to absorption. The output pulses are focused onto a Cd–Hg–Te detector D_1 which has a time constant of 5 nsec and operates at the temperature of liquid nitrogen (77°K). The detector D_2 mounted at the output of the monochromator M is used to monitor the frequency and intensity of the exciting pulses.

In order to vary the angle between the polarization directions of the exciting pulses in the experiments on the polarization properties of the photon echo, a quarter wave plate ($\lambda/4$) and a polarizer (P) were mounted in front of one of the lasers. For the quarter wave plate at 10.6 μ we used a thin film of Teflon which, apparently because of the way it is prepared, has a preferred direction and works as a uniaxial birefringent crystal. The $\lambda/4$ plate converts the linear polarization of the laser light to circular polarization and by rotating the polarizer (P) the polarization direction of one laser can be rotated about the other. The observation was made in two mutually perpendicular directions with the aid of an analyzer (P') located immediately in front of the detector D_1.

With this apparatus we have experimentally studied the dependence of the photon echo intensity in BCl_3 and SF_6 on the time between the exciting pulses, on their intensity, and on the concentration of gas particles, as well as the dependences of the polarization direction and intensity of the echo pulse on the angle between the polarization directions of the input pulses.

6. Measurements of the Relaxation Time T_2 in BCl_3 by the Photon Echo Technique. Measurement of T_1

In Section 4 it was shown that by studying the dependence of the photon echo pulse intensity on the time between the two exciting pulses it is possible to measure the irreversible collisional dephasing (the polarization relaxation time) of the medium, T_2. This assumes that the intensity of the photon echo is damped more as the interval τ_p is increased and with a characteristic time T_2 in accordance with Eq. (33).

Figure 12 shows some photographs of a series of input pulses taken with different time intervals between the exciting pulses. The third pulse is the photon echo pulse. Its intensity damps exponentially as τ_p is increased and, in the case of BCl_3 (Fig. 12a), is only a few percent of the exciting pulse intensities.

In SF_6 the photon effect is much stronger (Fig. 12b). Thus, in the upper photograph of this figure the intensity of the photon echo pulse of the P16 CO_2 laser line in SF_6 exceeds that of the first exciting pulse. In fact, these photographs were taken at a substantial pressure of the resonant gas, when the medium can hardly be considered optically thin, so the exciting pulses were absorbed by the gas (in this case the intensity of the first exciting pulse fell by roughly a factor of 4 on passing through the cell).

The quantity T_2 was measured in BCl_3 at a frequency of 951.16 cm^{-1} (the P12 CO_2 laser line) and in SF_6 at 945.94 cm^{-1} (P18). The results of the echo pulse intensity measurements are plotted as a function of $2\tau_p$ on a semilogarithmic scale in Fig. 13. The value of T_2 in BCl_3 as determined from the slope of a straight line fitted to the curve and reduced to a pressure of 1 Torr is 27 nsec • Torr. The same quantity measured in SF_6 was 24 nsec • Torr. The value of T_2 increases linearly with decreasing gas pressure (T_2, reduced to a pressure of 1 Torr, is constant). No deviation from linearity was observed over the entire working range. The accuracy in the measurement of T_2 is 20% and is determined mainly by the accuracy of the measurement of the gas pressure in the test cell using a manometer tube (type PMT-2) initially calibrated with a McLeod gauge.

It should be noted that the echo pulse amplitude is unstable even if the amplitude of the exciting pulses are ideally stable. This may be a result of the frequency instability of the

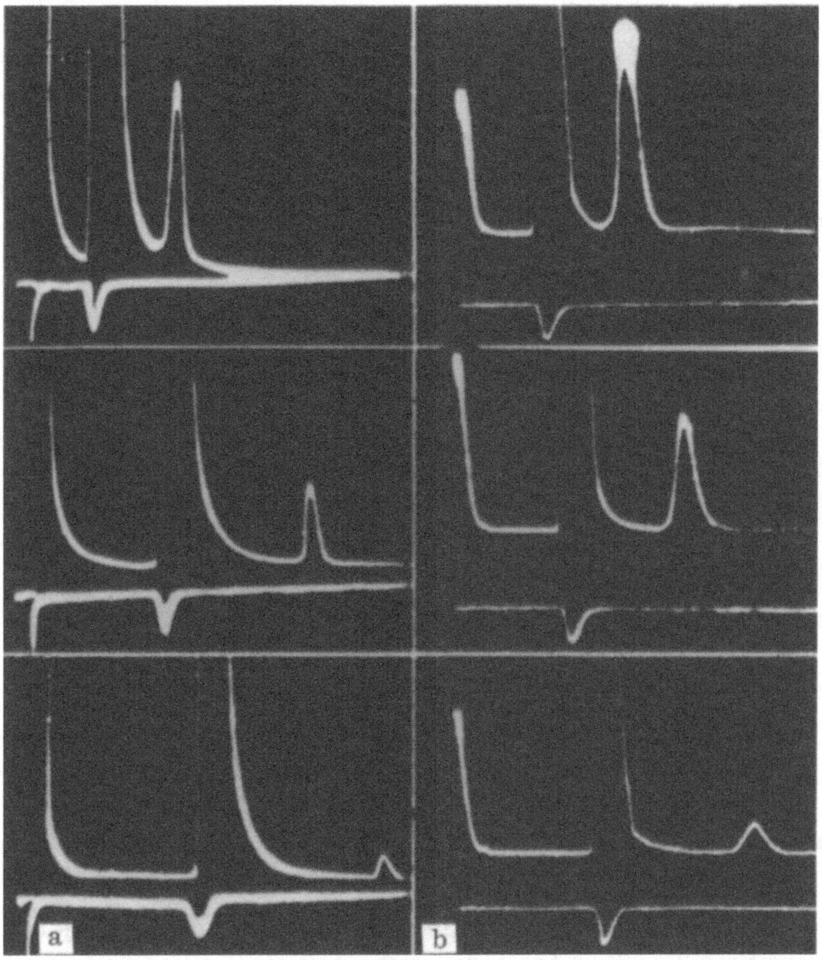

Fig. 12. Photographs of a series of input pulses with increasing τ_p. a) In BCl_3 gas; b) in SF_6. The third pulse is the photon echo pulse. The lower traces are the monitor channel pulses from D_2. The time scale is 1 μsec.

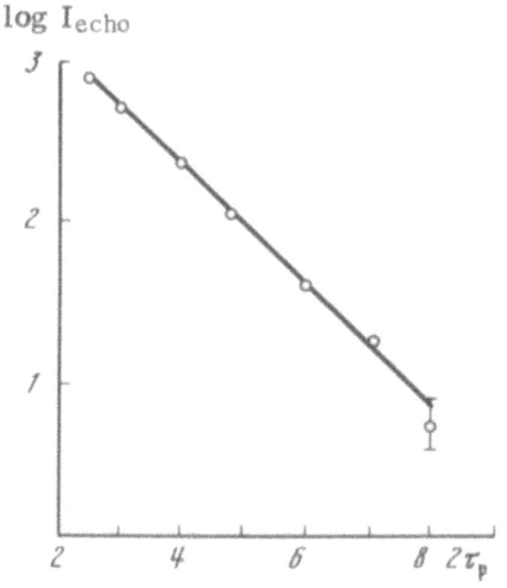

Fig. 13. A plot of the photon echo intensity in BCl_3 as a function of the time interval between the exciting pulses. The gas pressure is 5 mTorr.

lasers within the limits of the gain profile width; on the other hand, to use a single laser for forming both exciting pulses is experimentally difficult since it requires the production of microsecond delays. Thus, the measurements were made at the peak values of the intensities of the echo pulses, when the frequencies of the lasers and of the echo transition were the same. This is easily done thanks to the rather high repetition rate of the laser pulses (25-50 Hz).

Proceeding from the measured values of the time T_2, it is possible to estimate the effective cross section σ of the irreversible collisional dephasing process with the relation

$$v\sigma n = \nu = \frac{1}{2\pi T_2}, \tag{34}$$

where v is the most probable velocity of the gas molecules, n is the molecular concentration, and ν is the average frequency of dephasing collisions. Estimates of σ yield $\sigma_{BCl_3} = 9 \cdot 10^{-15}$ cm^2 and $\sigma_{SF_6} = 1.0 \cdot 10^{-14}$ cm^2. It is interesting that these values exceed the gas kinetic collisional cross sections for these gases $\sigma_{GK} = 5 \cdot 10^{-15}$ cm^2 by roughly a factor of 2. We estimated the value of the gas kinetic cross section σ_{GK} from the self-diffusion coefficient in SF_6, measured, for example, in [28]. In addition, the same values of SF_6 are obtained by proceeding from the known geometric dimensions of SF_6 and BCl_3 molecules [26]. Thus, phase shifting collisions are more probable than gas-kinetic collisions. This means that phase shift may take place, even in molecules which do not experience a gas kinetic collision, merely as a result of passing close enough to one another.

The experimental apparatus described in Section 5 can also be used to measure other relaxation times in molecular gases since the delay between the two laser pulses can be smoothly varied. Thus, for example, we have measured the relaxation time T_1 for the population difference in BCl_3 and SF_6 with this apparatus. The first pulse, at a high intensity, illuminated the medium and the second, of very much lower intensity, served as a test signal.

Following the illuminating pulse the difference in populations approaches the equilibrium value with relaxation time T_1. Then the absorption coefficient of the gas as a function of the delay time from the illumination pulse to the test pulse has the form

$$\alpha(\tau) = \alpha_0 (1 - e^{-\tau/T_1}), \tag{35}$$

where α_0 is the equilibrium absorption coefficient. The dependence of the test signal intensity on τ has the more complicated form of a double exponential,

$$I(\tau) = I_0 \exp\left\{-\alpha_0 L \left[1 - \exp\left(-\frac{\tau}{T_1}\right)\right]\right\} = I_\infty \exp\left[\alpha_0 L \exp\left(-\frac{\tau}{T_1}\right)\right], \tag{36}$$

where I_0 is the initial intensity of the test signal and $I_\infty = I_0 e^{-\alpha_0 L}$ is the intensity of the test signal in the absence of an illuminating pulse. Taking the logarithm of Eq. (36) twice we obtain

$$\ln \frac{\ln I(\tau) - \ln I_\infty}{\alpha_0 L} = -\frac{\tau}{T_1}. \tag{37}$$

Thus, by observing $I(\tau)$ experimentally and taking its logarithm twice as in Eq. (37) we obtain a value of the relaxation time T_1 from the slope of a linear fit. The measurements showed that the value of T_1 reduced to a pressure of 1 Torr for these gases is 35 ± 5 nsec · Torr, independently of the gas pressure in the range studied (5-50 Torr). The mechanism by which particles leave the excited state so rapidly is rotational relaxation which occurs at a high rate (due to the very dense rotational spectrum of BCl_3 and SF_6). The fact that T_1 is of the same order as T_2 must be taken into account when describing coherent interaction effects in CO_2 laser$-BCl_3$ or SF_6 systems.

Finally, it should be noted that T_1 can also be measured using photon echo techniques.

This conclusion follows from the close analogies between the photon and spin echo effects and from the fact that in his work on NMR Hahn measured the spin–lattice relaxation times T_1 in liquids by the secondary spin echo method. We have succeeded in observing for the first time the formation of a secondary photon echo in the optical range (Fig. 14). The secondary photon echo (fourth pulse) is the result of the action of the second exciting pulse and the primary photon echo pulse (third pulse) on the system.

Thus, with the experiments described in this section it has been possible to measure the relaxation time T_2 and, therefore, to measure the homogeneous widths ($\Delta\omega_{homog} = 2/T_2$) of absorbing lines in BCl_3 and SF_6 which are resonant with certain CO_2 laser lines, to estimate the cross sections for line broadening collisions, and to measure the longitudianl relaxation time T_1 in these gases.

7. The Polarization Properties of the Photon Echo

In the theoretical discussion of the photon echo effect in Section 4, it was pointed out that information could be derived about the type and number of the transition responsible for the formation of an echo by studying the dependences of the intensity and polarization direction of the photon echo on the intensity and angle between the polarization directions of the exciting pulses. Thus, in a study of the polarization properties of the photon echo in ruby [9] it was found that if the angle between the polarization directions of the first and second exciting pulses is ψ, then the polarization of the echo pulse is shifted in direction by an angle 2ψ relative to the first pulse. The type of transition responsible for the echo pulse in ruby was identified on this basis.

In their work [6] in which the photon echo effect was first observed in SF_6 at the P20 line of the CO_2 laser, Patel and Slusher found that the polarization direction of the echo coincides with the direction of the second exciting pulse, and from this they drew the conclusion, which seemed improbable to them, that the nondegenerate Q1 or P1 transition was responsible for the photon echo at this frequency.

In this section we shall present the results of a study of the polarization properties of the photon echo and of the dependence of the photon echo on the intensity of the exciting pulses in SF_6 at frequencies of 947.73, 945.94, and 944.15 cm^{-1}, corresponding to the P16, P18, and P20 CO_2 laser lines.

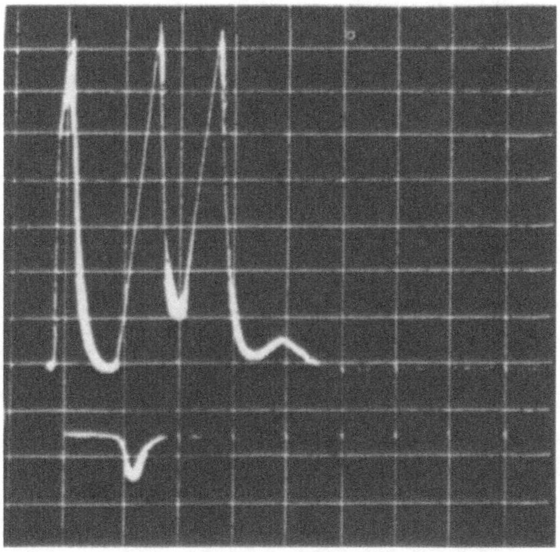

Fig. 14. An oscilloscope trace showing an observation of the secondary photon effect in SF_6 (P16).

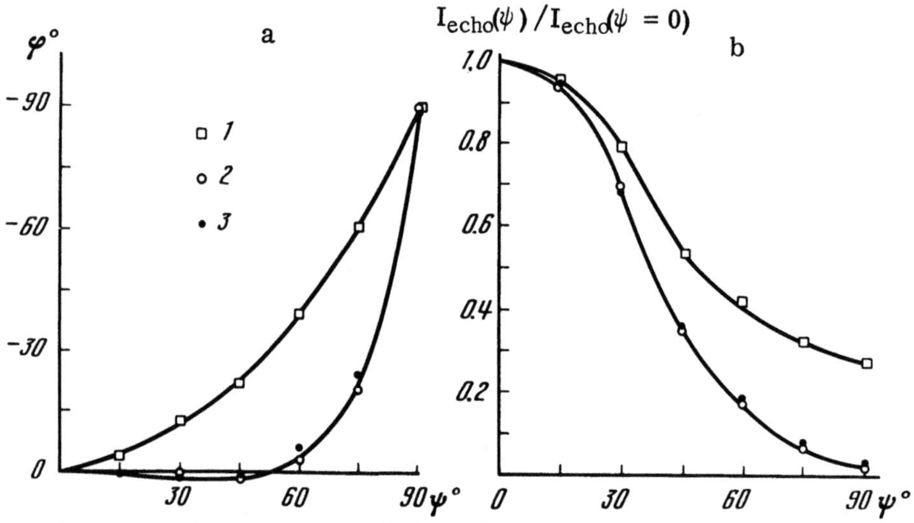

Fig. 15. Experimental plots of the polarization dependence of the photon echo in SF_6 at various CO_2 laser lines. a) Dependence of the angle φ on the angle between the polarization directions of the exciting pulses ψ; b) dependence of the photon echo intensity on ψ. 1) P16; 2) P20; 3) P18.

Figure 15 shows experimental plots of the polarization dependence of the photon echo in SF_6 at these frequencies. Also given there is the dependence of the angle φ between the polarization directions of the photon echo and the second exciting pulse on the angle ψ between the polarization directions of the two exciting pulses. Figure 15b shows the variation of the photon echo intensity with ψ. A comparison of the experimental curves with the theoretical curves of Fig. 9 shows that there is good agreement between the experimental curve 1 for the photon echo at 947.73 cm^{-1} (P16) and the theoretical curves for Q-branch transitions with sufficiently large J. The weak dependence of the behavior of the computed curves of Fig. 9 on J for $J > 1$ makes it impossible to estimate J from these plots.

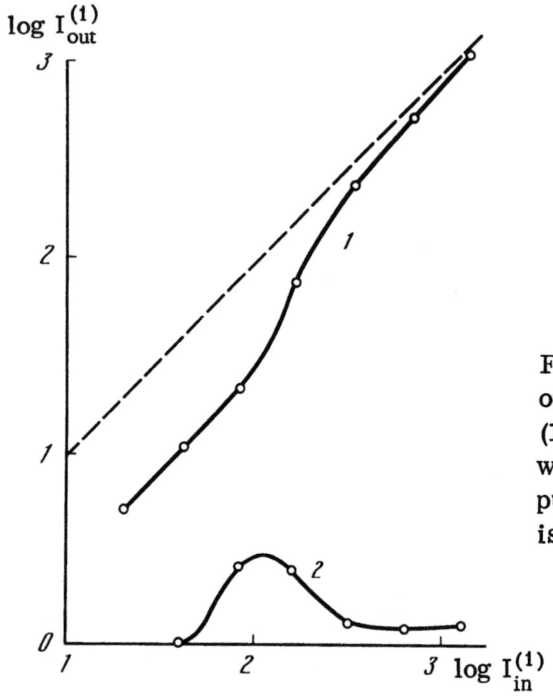

Fig. 16. The experimental variations of the output intensities of the first exciting pulse (1) and of the photon echo (2) in SF_6 (P16) with the input power of the first exciting pulse $I^{(1)}_{in}$. The intensity of the second pulse is $I^{(2)}_{in} = 4I^{(1)}_{in}$.

Some conclusion about the magnitude of J may be drawn from a study of the dependence of the photon echo intensity on the intensity of the exciting pulses (Fig. 16). In Fig. 16 the horizontal axis is the input intensity of the first exciting pulse (the intensity of the second pulse is four times greater) and the vertical axis is the output intensity of the first pulse (upper curve) and of the photon echo (lower curve). From the figure it is clear that the photon echo intensity peaks near the bend in the upper curve (that is, according to Section 3, near the $\pi/2$ region of the first exciting pulse) and falls off as the input intensity is increased further. The absence of repeated oscillations and the falling off of the photon echo intensity as the intensity of the input pulses is increased allows us to conclude (see Fig. 10) that the echo transition at 947.73 cm^{-1} (P16) is strongly degenerate and that J is large.

It should be noted that the photon echo, although it is of low intensity, is observed even when the intensity of the exciting pulses is substantially increased. This may be a consequence of an inhomogeneity in the exciting pulse beams across their transverse cross sections and the appearance of an echo in the periphery of the beams. Thus, for the curves of Fig. 16 the beam leaving the cell was extensively stopped down before focusing onto the detector. In addition, the existence of an echo signal at parameters considerably above optimal for $\pi/2$ and π exciting pulses may be explained by the shape of the absorbing line, whose Doppler width is much greater than the spectral width of the exciting pulses in our case. (In SF_6 and BCl_3 the Doppler width of the line $\Delta\omega_D = \frac{2\omega}{c}\sqrt{\frac{2kT \ln 2}{m}}$ is 31 and 35 MHz, respectively, at room temperature and 300 K, i.e., roughly 20 times greater than the spectral width of a 200-nsec pulse, $1/(\pi\tau_p) = 1.5$ MHz.) Thus, depending on how much the intensity of the exciting pulses is increased, the photon echo will be formed by particles which are farther from the line center, for which the exciting pulses would be $\pi/2$ and π pulses. Both these circumstances (we shall call them spatial and frequency inhomogeneity of the exciting pulses) were neglected in the theoretical discussion of the photon echo in Section 4 and explain the presence (not predicted there) of a photon echo even when the intensities of the exciting pulses are substantially increased.

The dependence of the photon echo intensity on the intensity of the exciting pulses at 944.15 cm^{-1} (P20) and 945.94 (P18) is similar to that described above for the photon echo at 947.73 cm^{-1}. However, the polarization dependence of the echo pulse is quite different. The polarization direction of the photon echo is practically the same as that of the second pulse, and only when the polarization of the exciting pulses are crossed ($\psi = 90°$) does the polarization of the echo become parallel to that of the first pulse ($\varphi = 90°$; see Fig. 15a), but its intensity is low, only 2-3% of the peak intensity (see Fig. 15b). This latter effect was evidently not noticed by Patel and Slusher [6], and from the fact that the polarization of the echo follows the polarization of the second pulse, the incorrect (we believe) conclusion was drawn in [11] that the well-resolved nondegenerate transition P1 or Q1 was responsible for the echo at 944.15 cm^{-1} (P20). To us this conclusion seems incorrect because it contradicts the falling off in the echo intensity with increasing intensity of the exciting pulses that was also noted by them. It is also apparently impossible to explain the observed polarization dependence by Rhodes' assumption [10] that the echo polarization must follow the polarization of the second pulse in view of the variation in the parameter θ of the exciting pulses over the length of the sample, since the polarization of the photon echo at 947.73 cm^{-1} (P16) behaves normally and, also, since a large artificial increase in the sample thickness does not change the way the photon echo polarization varies at this frequency.

The observed polarization dependence is, in our opinion, a result of the very complex structure of the absorbing transition. This is also indicated by the results of Lamb spectroscopy in SF_6 [22] which demonstrate the existence of several dips within the Doppler width of the absorption line. Thus, it should be kept in mind that several transitions which are close in frequency may simultaneously participate in forming the echo.

This experimental study of the polarization behavior and of the dependence of the photon echo intensity on the intensity of the exciting pulses has, therefore, led us to conclude that an absorbing transition in the Q-branch with a large value of J is responsible for the echo at 947.73 cm^{-1} (P16) in SF_6. The transitions which form echoes in SF_6 at 944.15 cm^{-1} (P20) and 945.94 cm^{-1} (P18) are also strongly degenerate; however, a more detailed theoretical analysis of the photon echo effect in gases is required to conclusively identify the types of these transitions.

8. The Concentration Dependence of the Photon Echo in SF_6

From the theory of spontaneous emission [13] it is known that if the linear dimensions of a system of molecules is small compared to the emitted wavelength, then it is impossible to regard the emission by each molecule as an independent act and the radiation from the whole system as the sum of elementary acts. In this case, for a certain method of exciting the system, the spontaneous emission intensity is given by

$$I_{sp}=\frac{\hbar\omega N^2}{4T_1^R}; \tag{38}$$

that is, it becomes proportional to the square of the number of radiating atoms, N. Here T_1^R is the radiative lifetime of an isolated molecule and $\hbar\omega$ is the energy of a photon.

Such a state of the system is known as "superradiant," and one of the ways of producing it is to send a $\pi/2$ pulse of coherent radiation (that is, a pulse with an envelope which obeys the condition $\frac{\mu}{\hbar}\int_{-\infty}^{+\infty}\mathcal{E}(t)\,dt=\frac{\pi}{2}$ and has duration τ_{pulse} much less than the transverse relaxation time T_2 of the medium) into a resonantly absorbing medium.

Let such a pulse be used to excite a system occupying a volume V with perpendicular cross section A and length l along the propagation direction V of the pulse such that all linear dimensions of the volume are very much greater than the wavelength of the light. This volume can always be divided into elementary volumes λ^3 provided there are at least a few excited particles in each, and then the contributions to the spontaneous emission from the elementary volumes are summed. The intensity of spontaneous emission from the entire volume is then

$$I_{sp}=\frac{\hbar\omega N^2}{4T_1^R}\int|\exp[i(\mathbf{k}-\mathbf{k}_z)\mathbf{r}_j]|^2_{avg}\,d\mathbf{k}. \tag{39}$$

The spontaneous emission from such a system, besides being proportional to the square of the number of particles, assumes a directionality and is emitted along the propagation direction of the exciting pulse $\mathbf{k}_z$. Integrating Eq. (39) yields [9]

$$I_{sp}=\frac{3\hbar\omega N^2\lambda^2}{16\pi T_1^R A}. \tag{40}$$

This expression is usually written in the form

$$I_{sp}=\frac{\hbar\omega N}{2T_s}, \tag{41}$$

where

$$T_s=\frac{8\pi T_1^R A}{3N\lambda^2}=\frac{8\pi T_1^R}{3\lambda^2 l n}. \tag{42}$$

The dependence of the spontaneous emission intensity on the square of the particle concentration $n = N/(Al)$ becomes understandable from this formulation. Spontaneous emission of N/2 particles takes place over a time T_2 which is inversely proportional to N, and this reduction in the radiative lifetime of the system is exactly due to the cooperative behavior of the radiators, when their phases are fully correlated. It also follows from Eq. (42) that, as opposed to a system with dimensions less than the wavelength (in which the time $T_s = 2T_1^R/N$ is determined by the total number of excited particles N/2), in a macroscopic system ($V \gg \lambda^3$) the time T_s is determined by the number of particles $n\ (3\lambda^2 l/8\pi) \ll N$ inside an effective coherent volume which is essentially the volume of the diffraction cone.

Inhomogeneous broadening of the excited transition leads to complete dephasing of the radiators over a time $T_2^* = 2\sqrt{\ln 2}/\Delta\omega_D$, where $\Delta\omega_D$ is the inhomogeneous broadening of the absorption line, due to the Doppler effect in our case. As already noted the important feature of this dephasing is its reversibility, which is realized by sending a second exciting π pulse into the medium delayed relative to the first by a time τ_p. The photon echo signal at time $2\tau_p$ is the result of the superradiance of the once-again phased ensemble.

In describing the photon echo effect it is customary to neglect the radiation of the system immediately after passage of the first exciting pulse (which produces the superradiant state) by assuming that $T_2^* \ll T_s$, i.e., that the very much more rapid Doppler dephasing prevents radiative decay of the superradiant state. The possibility of decay of the superradiant state immediately after passage of the first exciting pulse was excluded from the considerations of Section 4 by neglecting the secondary effect on the polarization due to the field produced by the polarization of the medium.

Since the time T_s decreases proportionally to n, as the particle concentration is increased the radiative decay of the superradiant state immediately after passage of the first exciting pulse becomes even more important, thereby leading to destruction of the quadratic concentration dependence of the photon echo intensity [23]. We have investigated this dependence experimentally for the photon echo in a CO_2 laser–SF_6 system at the P18 laser line. Use of a gaseous medium for studies of the concentration dependence make it possible to smoothly vary the particle concentration within wide limits, thereby greatly simplifying and increasing the reliability of the experiment compared to the analogous experiment in ruby [24].

The photon echo pulse was produced by the apparatus described in Section 5 in a cell of length 3 m after passage of two exciting CO_2 laser pulses of duration $\tau_{pulse} = 150$ nsec separated by an interval of $\tau_p = 500$ nsec. The SF_6 gas pressure in the cell was varied over wide limits and measured with a PMT-2 tube.

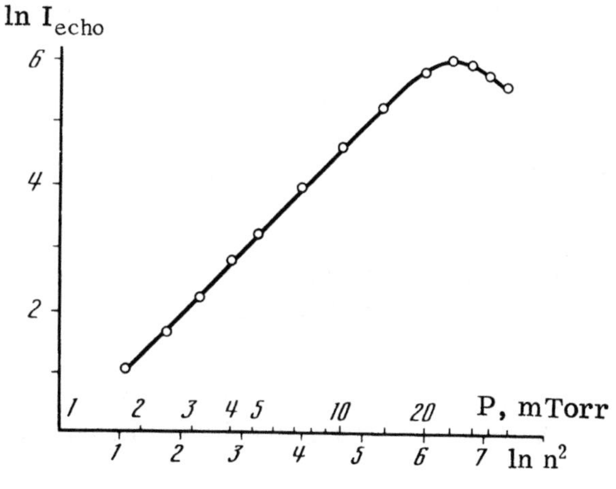

Fig. 17. A plot of the concentration dependence of the photon echo intensity.

Figure 17 shows the dependence of the photon echo pulse intensity on the square of the gas concentration on a logarithmic scale. The fact that the photon echo intensity decreases in proportion to $e^{-4\tau_p/T_2}$ due to irreversible collisional dephasing over a characteristic time T_2 as the pressure is increased was taken into account in constructing the curve. In order to increase the accuracy of the experiment, the value of T_2 was measured at each working point using the technique described in Section 6, and the photon echo intensity was extrapolated to the value at $\tau = 0$ by multiplying by $e^{4\tau_p/T_2}$.

It is clear from an analysis of Fig. 17 that as the gas concentration is increased the photon echo intensity increases with n^2, but later, at a pressure of 25 mTorr, the curve bends.

As pointed out above, the location of the bend corresponds to a particle concentration at which the time T_s becomes of order T_2^*. However, under our experimental conditions $T_2^* \ll \tau_{pulse}$; thus, only part of the Doppler profile of the absorption line, determined mainly by the spectral width of the exciting pulse $1/\tau_{pulse}$, is involved in the formation of the echo. Thus, the bend in the curve of Fig. 17 must correspond to $T \approx \tau_{pulse}$.

The radiative decay time T_s of the superradiant state near the bend in the curve may be estimated from Eq. (42). Then it is necessary to include the fact that, due to the Boltzmann distribution of particles over the rotational sublevels in the vibrational ground state of the SF_6 molecule, only a fraction $1.8 \cdot 10^{-4}$ of the total number of particles can be within the Doppler profile of the absorbing line (see Section 11). In addition, only that part of the Doppler profile given by the condition $\sim T_2^*/\tau_{pulse}$ takes part in the formation of the echo. Thus, the concentration of active particles near the bend in the curve (i.e., at 25 mTorr) is $n \approx 10^{10}$ cm^{-3}. An estimate of the spontaneous lifetime of an isolated molecule, T_1, using the formula

$$T_1^R = \frac{3\hbar\lambda^3}{32\pi^3\mu^2}, \tag{43}$$

yields $T_1^R = 4.7 \cdot 10^{-2}$ sec if we begin by taking the matrix element of the dipole moment of the transition in Eq. (43) to be $\mu = 2.8 \cdot 10^{-19}$ cgse (see Section 11). Substituting these values of n and T_1^R into Eq. (42) yields an estimate of T_s in the region where the quadratic dependence is destroyed: $T_s \approx 130$ nsec ($T_s \approx 4$ nsec · Torr). This is practically the same as the length of the exciting pulses and, therefore, experimentally confirms the fact that the quadratic concentration dependence of the photon echo intensity is destroyed because of the decay of the superradiant state of the system immediately after passage of the first exciting pulse. The reduction in the lifetime from $T_1^R \approx 4.7 \cdot 10^{-2}$ sec to $T_s \approx 1.3 \cdot 10^{-7}$ sec caused by spontaneous emission from the molecules is a consequence of the cooperative behavior of the radiators.

Therefore, by experimentally studying the concentration dependence of the photon echo we can isolate the range of pressures in which the decay of the superradiant state immediately after passage of the first exciting pulse becomes important and evaluate the radiative lifetime T_s of the superradiant state due to spontaneous emission.

CHAPTER III

THE NUTATION EFFECT IN BCl_3 AND SF_6

This chapter deals with a study of the optical nutation effect on vibrational–rotational absorbing transitions in BCl_3 and SF_6 resonant with the P12, P16, P18, and P20 CO_2 laser lines [27].

Nutation was observed against a constant background of CO_2 laser light following illumination of the medium by a short, high-intensity pulse at the same frequency [29].

Optical nutation is a coherent interaction between a field and matter and involves a modulation in the amplitude of a light field passing through a resonant medium. The modulation frequency, known as the nutation frequency, is proportional to the amplitude of the field and to the dipole moment of the resonant transition. The perturbing field must be turned on in a time shorter than the nutation period, and the fundamental condition for a coherent interaction must also be satisfied (that is, the nutation period must be much shorter than the transverse relaxation time T_2 of the medium).

An experimental study of nutation makes it possible, by measuring the period of the oscillations and measuring the electric field strength of the light wave, to determine the dipole moments of the transitions responsible for producing the nutation. The examination of nutation in molecular gases in the infrared is particularly interesting because it is essentially the only reliable way to measure the dipole moments of vibrational–rotational transitions in gases. The determination of the dipole moments of these transitions by measuring the radiative lifetimes of vibrational–rotational transitions is extremely difficult because these times are, as a rule, of the order of seconds. Measuring the dipole moments of vibrational–rotational transitions via the absorption coefficient requires a determination of the total number of particles involved in the absorption since, because of the distribution of the particles over the rotational sublevels of the ground vibrational state, only those particles lying in one of these rotational sublevels are involved in absorption by the given vibrational–rotational transition. This fraction of particles q_0 can only be estimated fairly crudely theoretically beginning with a Boltzmann distribution of particles over the rotational sublevels of the ground vibational state. In addition, measuring the dipole moments using the nutation effect and measuring the absorption coefficients of the corresponding transitions make it possible to determine experimentally the value of q_0, knowledge of which is often required for applications.

9. Theoretical Description

In the theoretical description of nutation, as in the case of self-induced transparency, a semiclassical approximation is used in which an ensemble of particles is described quantum mechanically with the aid of the density matrix, the polarization of the medium induced by the field is then calculated, and the reverse effect of the polarization on the perturbing field is described classically by a wave equation.

We shall assume that a perturbation in the form of a step-function electric field from a light wave of the form

$$\mathbf{E}(z,\ t)=\begin{cases} \mathbf{x}\mathscr{E}(z,\ t)\cos(\omega t-kz), & t\geqslant 0,\ \ z>0,\\ 0, & t<0,\ \ z>0, \end{cases} \tag{44}$$

where $\mathscr{E}$ is the amplitude and the frequency ω is close to the transition frequency ω_0, is incident on a medium which may be regarded as a two-level system. The behavior of a two-level system with such a perturbation is described by Eq. (9).

For exact resonance $\Delta = 0$, and neglecting the secondary effect on the polarization due to the field produced by the polarization itself, that is, in the approximation of a given field,

$$\mathscr{E}(z,\ t)=\begin{cases} \mathscr{E}_0, & t\geqslant 0,\ \ z>0,\\ 0, & t<0,\ \ z>0, \end{cases}$$

Eq. (9) reduces to an equation for the absorbing component of the polarization V responsible for the appearance of nutation,

$$\frac{d^2V}{dt^2}+\left(\frac{1}{T_1}+\frac{1}{T_2}\right)\frac{dV}{dt}+\left(\frac{\mu^2\mathscr{E}_0^2}{\hbar^2}+\frac{1}{T_1T_2}\right)V=0, \tag{45}$$

whose solution has the form

$$V(t)=n\mu\sin\Omega t\cdot\exp\left[-\frac{t}{2}\left(\frac{1}{T_1}+\frac{1}{T_2}\right)\right], \tag{46}$$

where

$$\Omega=\left[\frac{\mu^2\mathcal{E}_0^2}{\hbar^2}-\frac{1}{4}\left(\frac{1}{T_2}-\frac{1}{T_1}\right)^2\right]^{1/2}. \tag{47}$$

Thus, the polarization of the medium varies harmonically in time with a frequency Ω determined mainly by the magnitude of the electric field and the dipole moment of the transition, and the amplitude of the oscillations decays exponentially with a time constant $T_2' = 2T_1T_2/(T_1+T_2)$.

The dependence of Ω on the relaxation time becomes important when the nutation period is comparable to T_2 under the usual assumption $T_2 \ll T_1$. However, as our measurements showed (cf. Section 6), in BCl_3 and SF_6 the time T_1 differs little from T_2. This is an indication that the probability of elastic collisions (i.e., those which shift the phase of radiating molecules without changing their energy) is low. It is thus possible to neglect the dependence of the nutation frequency on the relaxation times $1/T_1-1/T_1$ even when the nutation period is comparable to T_2 and, furthermore, to assume that both the polarization and the population difference relax with the same time T_2. Under these conditions the solution (46) of Eq. (45) takes the form

$$V(t)=n\mu\sin\frac{\mu\mathcal{E}_0}{\hbar}t\cdot\exp\left(-\frac{t}{T_2}\right). \tag{48}$$

Since the polarization of the medium is related to the field by the wave equation (12), which has the form (13) for slowly varying amplitudes, the field which passes through the medium is given by

$$\mathcal{E}=\mathcal{E}_0\left[1-\frac{\alpha z}{\Omega T_2}\sin\Omega t\cdot\exp\left(-\frac{t}{T_2}\right)\right], \tag{49}$$

This is obtained after substituting the polarization (48) into the wave equation (13) and integrating subject to the condition $\omega \gg \Omega$. Thus this field is modulated at frequency $\Omega=\mu\mathcal{E}_0/\hbar$. The modulation depth is determined by the gain coefficient $\alpha/\Omega T_2$ and the sample length z. Here $\alpha = 4\pi n\omega T_2\mu^2/\hbar c$ is the linear absorption coefficient. The field oscillations are damped with characteristic time T_2.

Solutions (48) and (49) are valid for a nondegenerate two-level system with homogeneous broadening. Real systems, molecular gases, are generally strongly degenerate in the projection (m) of the rotational angular momentum (J) along a given direction and are also inhomogeneously Doppler broadened.

In order to include the effect of degeneracy on nutation, it is necessary to sum the contributions to the polarization of the medium due to the different degenerate states. The dependence on m and J of the dipole moment for transitions between degenerate states acted on by linearly polarized light for symmetric-top molecules is determined by Eqs. (19) and (20). Assuming that all degenerate states are equally probable and, therefore, equally populated with density $n_m = n/(2J+1)$, the polarization per unit volume of a degenerate medium is given by

$$V(t)=\sum_{m=-J}^{J}\frac{n\mu_m}{2J+1}\sin\frac{\mathcal{E}_0\mu_m}{\hbar}t. \tag{50}$$

We shall again consider the case of large angular momenta $J \gg 1$ and go to the quasiclassical description assuming that the distribution of dipole moments over the various degenerate states is continuous, that is, assuming that $\mu = \mu_0 \cos\beta$ for the Q-branch and that $\mu = \mu_0 \sin\beta$ for the P- and R-branches, where $\cos\beta = m/J$ and $\sin\beta = (J^2 - m^2)^{1/2}/J$. In this case, going from summation to integration over all possible orientations of the angular momentum, we obtain expressions for the polarization per unit volume of a degenerate medium:

$$V^Q(t) = n\mu_0\left(\frac{\sin\Omega t}{\Omega^2 t^2} - \frac{\cos\Omega t}{\Omega t}\right), \tag{51}$$

$$V^{P,R}(t) = n\mu_0 \int_0^{\pi} \sin^2\beta \cdot \sin(\Omega t \sin\beta)\, d\beta. \tag{52}$$

The last integral, although it is derived analytically, is written in the form of a series of Bessel functions and integrated numerically. Figure 18 shows plots of the dependence of the polarization per unit volume of gas on the time for a nondegenerate system with dipole moment μ_0 [see Eq. (48)] and for a degenerate system of Q-branch (51) and P- and R-branch (52) transitions with a maximal dipole moment μ_0.

It is important that in a degenerate system nutations are damped even in the absence of relaxation, with nutations in the Q-branch transition being damped much more than those in the P- or R-branches. This fact may be used to identify the type of transition responsible for the nutations.

From the graphs of Fig. 18 it is also clear that degeneracy leads to a change in the nutation frequency. Then a degenerate system can always be placed in a correspondence with a damped nondegenerate system having an effective dipole moment μ_{eff} related to μ_0 by some equations which follow from the condition that the nutation frequencies in an effectively nondegenerate system with dipole moment μ_{eff} and a real degenerate system with maximal dipole moment μ_0 coincide, i.e.,

$$\mu_{eff}^Q = \frac{\mu_0}{1.41}; \qquad \mu_{eff}^{PR} = \frac{\mu_0}{1.14}.$$

It is this effective dipole moment which is measured experimentally and determines such parameters of the gas as its absorption coefficient, absorption cross section, and so on.

We shall take inhomogeneous broadening of the transition due to the Doppler effect into account for the case in which the nutation frequency is very much less than the inhomogeneous linewidth, since with our technique it is just this case which is observed.

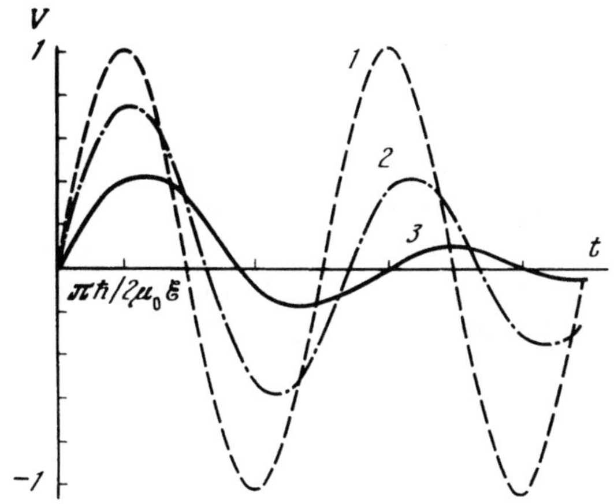

Fig. 18. Dependence of the absorbing component of the polarization of the medium on the time for a nondegenerate system with dipole moment μ_0 (1) and for a degenerate system with maximal dipole moment μ_0 ($J \gg 1$): Q-branch transition (2); P- and R-branch transitions (3).

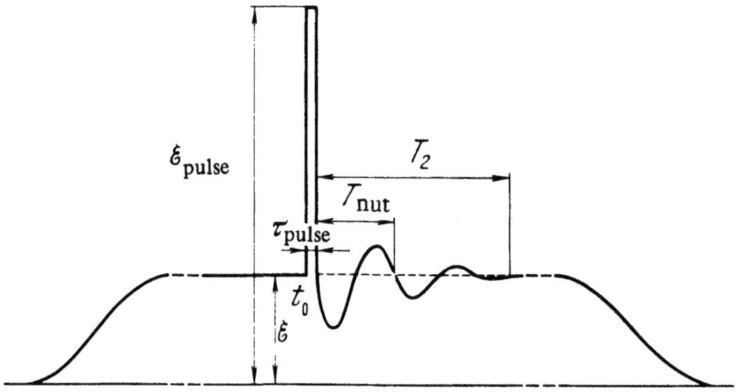

Fig. 19. The scheme for observing nutation at the level of a constant perturbation $\mathcal{E}$ immediately after irradiation of the medium at time t_0 by a short pulse with amplitude $\mathcal{E}_{pulse}$ and duration τ_{pulse} ($\tau_{pulse} \ll T_{nut}$).

Since nutation is observed against a background of constant-power radiation passing through the medium immediately after a short pulse of much greater power at the same frequency, it is necessary to solve Eqs. (9) for the absorption component of the polarization $V(\Delta, t)$ as a function of the frequency difference and time with initial conditions specified at time $t_0 + \tau_{pulse}$ (Fig. 19). The polarization of the medium at time t_0 is determined by the stationary solutions of Eqs. (9):

$$V(t_0, \Delta) = \frac{n\mu}{T_2 \Omega} \frac{\Omega^2}{\Omega^2 + \Delta^2}. \tag{53}$$

The polarization at time $t_0 + \tau_{pulse}$ (i.e., after the pulse of length τ_{pulse} and amplitude $\mathcal{E}_{pulse}$) is given by

$$V(\Delta, t_0 + \tau_{pulse}) = V(\Delta, t_0) \cos\theta, \tag{54}$$

where $\theta = \mu \mathcal{E}_{pulse} \tau_{pulse}/\hbar$ and the condition $\mathcal{E}_{pulse} \gg \mathcal{E}$ is used. The solution of Eqs. (9) for $V(\Delta, t)$ as a function of the frequency difference and time with the initial conditions (54) has the form

$$V(\Delta, t_0 + \tau_{pulse} + t) = V(\Delta, t_0 + \tau_{pulse}) \cos\sqrt{\Omega^2 + \Delta^2}\, t. \tag{55}$$

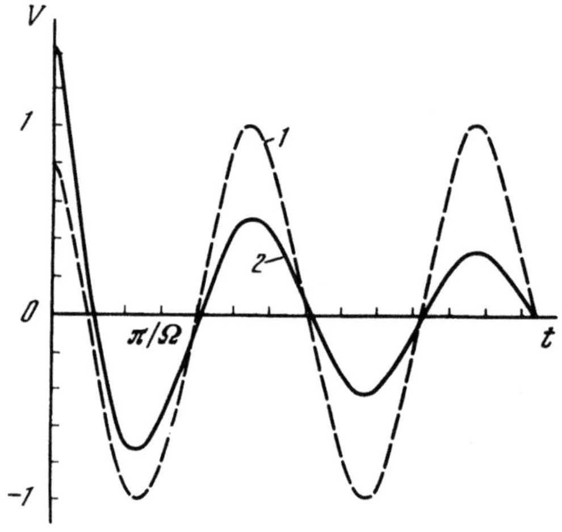

Fig. 20. The time dependence of the absorption component V of the polarization of the medium: 1) Inhomogeneous broadening of the line and $\Delta\omega_D \gg \Omega$; 2) homogeneous broadening with $\Delta\omega_{homog} = 2/T_2 \ll \Omega$.

To find the total polarization it is necessary to integrate this equation over all frequencies Δ, i.e.,

$$V(t) \sim \int_{-\infty}^{\infty} \frac{\Omega^2}{\Omega^2 + \Delta^2} \cos \sqrt{\Omega^2 + \Delta^2}\, t d\Delta. \quad (56)$$

Equation (56) was integrated numerically and the results are shown in Fig. 20, from which it is apparent that inhomogeneous broadening also leads to damping of the nutation.

An experimental study of nutation thus makes it possible to measure the effective dipole moment of the absorbing transition when the nutation frequency and electric field strength of the light wave are known. In addition, the type of the absorbing transition may be determined from the difference in the way the nutation is damped for Q-, P-, and R-branch transitions. Then it should be kept in mind that there may be three reasons for nutation damping: relaxation, degeneracy, and inhomogeneous broadening.

10. Experimental Apparatus

The optimum shape of the perturbing field for observing nutation is a perturbation in the form of a step-function electric field of the light wave. Strict limitations are then placed on the risetime of the step: $\tau_r < T_{nut} < T_2$. There are considerable technical difficulties in the way of producing such a perturbation.

Hocker and Tang [7], who first observed optical nutation, used a pulse from a Q-switched CO_2 laser of length 300 nsec with a risetime of 100 nsec as a perturbing field. Then, although it was possible to observe the effect, it was extremely difficult to measure the dipole moment of the resonant transition because the field was not constant in time and because of absorption of the pulse by the medium. In addition, the pulse risetime was comparable to the nutation period.

Thus, we have used a scheme for observing nutation in our studies. Nutation was observed against a constant background power level of CO_2 laser light following irradiation of the medium by a short, high-power pulse at the same frequency and resonant with the transition frequency (see Fig. 19). Then the pulse serves as a rapidly rising front for a constant perturbation, and this scheme for observing nutation approaches the optimum, in which the perturbation is a step-function electric field from the light wave.

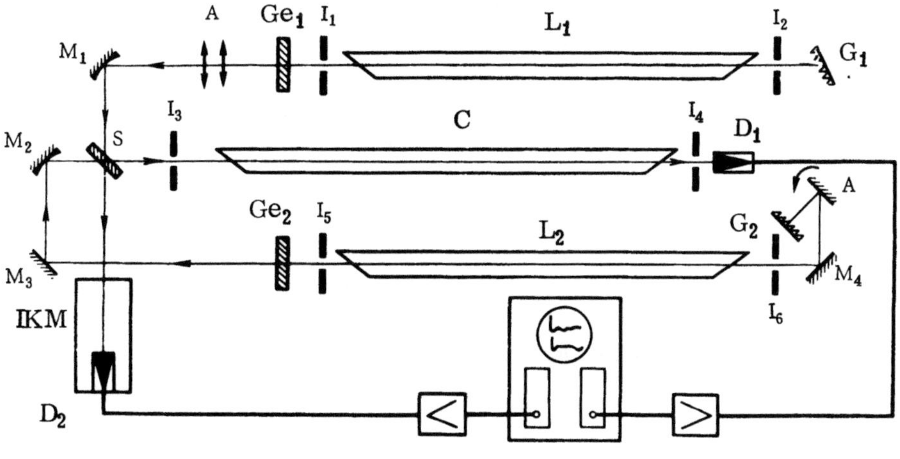

Fig. 21. A diagram of the experimental apparatus for studying the nutation effect in SF_6 and BCl_3.

According to this scheme the perturbation is produced by two CO_2 lasers arranged as shown in Fig. 21. The constant perturbation is provided by laser L_1 which operates in a quasi-continuous regime (that is, it is modulated by the supply to the laser discharge tube) with a pulse of about 1 msec in length at an intensity of several W/cm^2. The laser cavity is made up of a dielectric-coated germanium plate with a reflectivity of 70% for the output mirror and a diffraction grating G_1 (100 lines/mm). Modulating the current in the laser tube greatly increases the gain of the active medium, ensures high output stability, and makes it possible to work with a low-Q cavity.

In the second laser L_2, the cavity Q is modulated (Q-switched) with a rotating mirror R in addition to being modulated by the supply. The cavity of this laser is formed by a semi-transparent germanium mirror Ge_2, the rotating mirror R, and a diffraction grating G_2. This laser yields pulses of length 100-200 nsec and intensity 100 W/cm^2 which serve as a short front for the constant perturbation.

The use of diffraction gratings makes it possible to tune the laser output frequencies over a wide range of laser lines. The output frequency and intensity are monitored with an IKM-21 monochromator which has a diffraction grating to increase its resolution.

The output of laser L_1 is attenuated by an attenuator A made up of two polarizers, and both beams are directed into the cell C (length 3 m) containing the test gas with the aid of spherical steering mirrors M_1-M_3 and a beam splitter S. The use of spherical mirrors makes it possible to compensate for the diffractive divergence of the beams, and, therefore, ensures constant radiant intensity at the center of the beam over the entire sample. The light is detected by a Cd–Hg–Te detector with a time constant of 5 nsec and a detector area of $0.4 \cdot 0.4$ mm^2. Thus, a beam with a diameter of 5-10 mm has negligible inhomogeneity over its perpendicular cross section.

Special attention was paid to calibration of the detector. Calibration was done in two ways: by ordinary calorimetry of millisecond pulses with a measurement cone and by using a cw CO_2 laser whose power is measured by the evaporation of liquid nitrogen due to absorption of the laser beam and then greatly attenuating the beam with a porous shutter and letting it strike the detector. The agreement between the two methods is an indication of the reliability of the results.

11. Discussion of the Experimental Results

This apparatus was used to observe nutation in SF_6 at frequencies of 947.73, 945.94, and 944.15 cm^{-1}, which correspond to the P16, P18, and P20 CO_2 laser lines, and in BCl_3 at 951.16 cm^{-1} (P12). Typical oscilloscope traces of the effect are shown in Fig. 23. The origin in the traces is the short pulse from laser L_2. Then some oscillations (nutation) about the constant

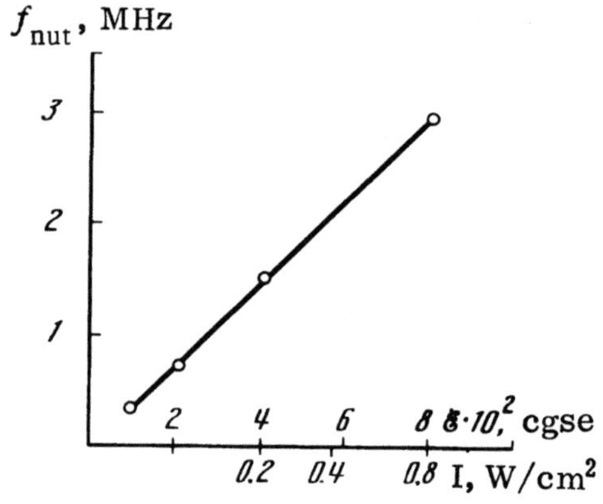

Fig. 22. A plot of the dependence of the nutation frequency on the field (intensity) of the cw laser.

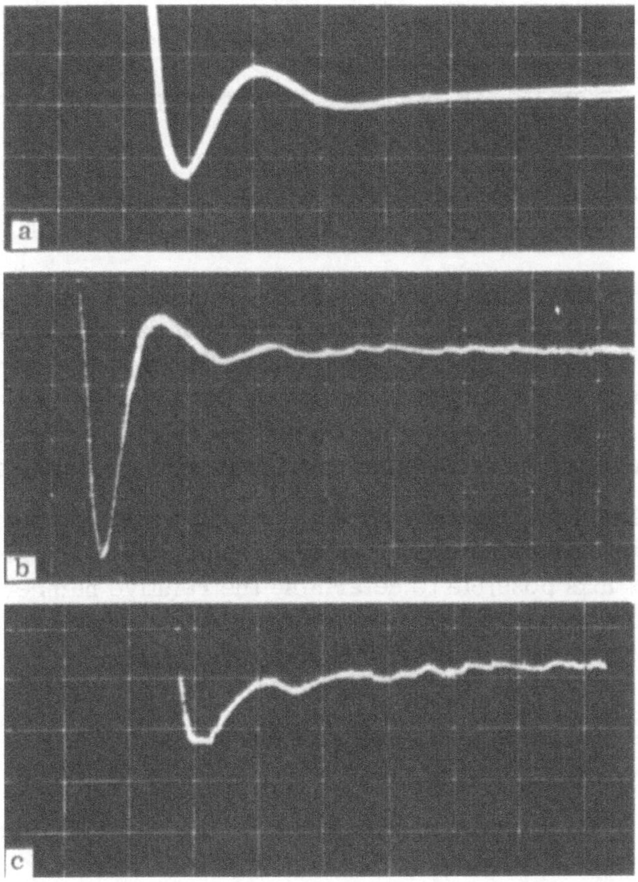

Fig. 23. Some oscilloscope traces of nutation. In SF_6: a) at 947.73 cm^{-1} (P16), time scale 0.25 μsec; b) at 944.14 cm^{-1} (P20), time scale 0.5 μsec; in both cases $T_2 = 1.5$ μsec. In BCl_3: c) at 951.16 cm^{-1} (P12), time scale 0.5 μsec.

power level of laser L_1 follow, whose frequencies depend on the intensity of laser L_1 and are independent of the gas pressure.

The nutation period and the intensity of the field of the quasicontinuous laser (L_1) were measured. The gas pressure in the cuvette lay within 5-50 mTorr and was set so that this laser field saturated the transition and passed through the medium practically without absorption (usually the output intensity was 90% of the input) since a change in the field along the sample due to absorption would introduce an uncertainty in the dipole moment measurement. In addition, simultaneous monitoring of the field intensity and gas pressure makes it possible to ensure satisfaction of the inequality $\Omega < 1/T_2$. Since when the radiant intensity is reduced the nutation frequency is reduced, it is then necessary to reduce the absorbing gas pressure in order to saturate the transition; however, this causes a lengthening of the relaxation time T_2.

Figure 22 shows a graph of the dependence of the nutation frequency on the field intensity. The linearity of this curve allows us to use the relation $f_{nut} = \Omega/2\pi = \mu_{eff}\,\mathcal{E}_0/2\pi\hbar$ to determine the dipole moment. It should be noted that this linearity is destroyed if the perturbing field is increased further, and the effect disappears since the nutation period approaches the pulse length τ_{pulse} which serves as the rising front of the constant perturbation, and the nutation frequency is limited by the spectral width of the pulse.

TABLE 1

Gas	Frequency, cm^{-1}	Laser line	α, $cm^{-1} \cdot Torr^{-1}$	$\mu_{eff} \cdot 10^{-19}$, cgse	$\sigma \cdot 10^{13}$, $cm^2 \cdot Torr$	$q \cdot 10^4$, $Torr^{-1}$	$q_0 \cdot 10^4$
BCl_3	951.16	*P12*	0.13	2	0.8	0.5	0.8
SF_6	944.15	*P20*	0.38	2.5	1.0	1.1	1.6
SF_6	945.94	*P18*	0.55	2.6	1.3	1.2	1.8
SF_6	947.73	*P16*	1.05	2.8	1.3	2.4	3.6

The dipole moments determined with the above technique were $\mu_{eff} = 2.8 \cdot 10^{-19}$, $2.8 \cdot 10^{-19}$, and $2.5 \cdot 10^{-19}$ cgse in SF_6 for the transitions resonant with the P16, P18, and P20 CO_2 laser lines and $\mu_{eff} = 2.0 \cdot 10^{-19}$ cgse for the absorbing transition in BCl_3 resonant with the P12 line. The accuracy of these values is mainly determined by the accuracy of the detector calibration and is 10% at best. The relative accuracy is determined only by the accuracy with which the period of the observed nutation was measured and is therefore considerably greater.

The linear absorption coefficients α of SF_6 and BCl_3 were also measured at these frequencies. The values of α reduced to the absorbing gas pressure are given in Table 1. From these values of μ_{eff} and α it is possible to determine the relative number of particles q_0 in the lower level of the resonant transition, that is, the fraction of particles lying within the Doppler profile of the resonant absorbing transition, by using the relation

$$\alpha = \frac{4\pi^2 n_0 q_0 \omega \mu_{eff}^2}{\hbar c \Delta\omega_D}, \tag{57}$$

where $\Delta\omega_D$ is the width of the Doppler profile and n_0 is the concentration of all gas particles. The values of q_0 for the transitions studied here are given in Table 1.

This quantity may also be evaluated, starting with a Boltzmann distribution of molecules over the rotational sublevels of the vibrational ground state, if the molecular configurations and the rotational constant are known. For symmetric-top molecules the relative number of particles in a given rotational state is given by

$$q_T = \frac{(2J+1)\exp\{-[BJ(J+1)+(C-B)K^2]/kT\}}{\sum\limits_{J=0}^{\infty}\sum\limits_{K=-J}^{J}(2J+1)\exp\{-[BJ(J+1)+(C-B)K^2]/kT\}}. \tag{58}$$

The statistical rotational sum in the denominator of this equation is approximated by $\sum\limits_J \sum\limits_K \approx \sqrt{\frac{\pi (kT)^3}{B^2 c}}$ in the case of a dense spectrum ($B \ll kT$) [25]. The numerator as a function of J and K has an extremum at $J \approx 32$ (in SF_6) and at $J = K \approx 46$ (in BCl_3) if we start with the rotational constants of the BCl_3 ($B = 2C \approx 0.1$ cm^{-1}) and SF_6 ($B \approx C \approx 0.1$ cm^{-1}) molecules [26]. Estimates of the maximum q_T^{max} yield $q_T^{max} \approx 2 \cdot 10^{-4}$ for both gases.

In estimating q_T^{max} for the SF_6 molecule, which is a spherical-top molecule, it was assumed that the degeneracy in K characteristic of this type of molecule had been removed. The agreement between q_0 and q_T^{max} as to order of magnitude confirms this assumption experimentally.

In applications it is often necessary to know the relative number of particles q which take part in absorption. At low pressures, when the homogeneous linewidth does not exceed the Doppler width ($2/T_2 = \Delta\omega_{homog} \ll \Delta\omega_D$), q is less than q_0 since only part of the molecules in the Doppler profile, specified by the ratio $\Delta\omega_{homog}/\Delta\omega_D$, are involved in absorption. As the pressure is increased and the homogeneous width of the absorbing line becomes greater than the Doppler width, in SF_6 and BCl_3, which have extremely dense rotational spectra, q becomes greater than q_0 and continues to increase due to the inclusion of neighboring transitions in the absorption.

The values of q for the transition studied here were evaluated (see table) using the equation

$$\alpha = \frac{4\pi\omega n_0 q T_2 \mu^2_{\text{eff}}}{\hbar c}, \tag{59}$$

where the time T_2 was found from experiments on the photon echo (see Section 6). Equation (58) was also used to determine the absorption cross section of the homogeneous lines $\sigma = \alpha/n_0 q = 4\pi\omega T_2 \mu^2_{\text{eff}}/\hbar$ at these frequencies. These cross sections decrease with rising pressure. The values of σ reduced to a pressure of 1 Torr are given in the table.

It should be kept in mind that these values of q and σ were measured at low pressures, 5-50 mTorr, for specific absorbing transitions. Extrapolation of these values to higher pressures is generally arbitrary, since it requires that the dipole moments of the transitions which are newly added to the absorption profile as the pressure is increased be of the same order of magnitude as those of the transitions already studied. Nevertheless, the values obtained are a completely satisfactory estimate of these quantities up to pressures of several Torr. This is indicated, for example, by the constancy of the absorption coefficient α, referred to the pressure, over this pressure range.

In the theoretical discussion of nutation it was noted that it is possible to draw conclusions about the type of transition responsible for the nutation on the basis of differences in the way the nutation is damped. Figure 23a and b are oscilloscope traces of this effect in SF_6 at 947.73 cm^{-1} (P16) and 944.15 cm^{-1} (P20), respectively. The time T_2 and the broadening inhomogeneity are the same in both cases. The more rapid nutation damping at 947.73 cm^{-1} (P16) is apparently an indication that a Q-branch transition is responsible for absorption by SF_6 at this frequency and that P- or R-branch transitions are responsible for absorption at 944.15 cm^{-1} (P20) and 945.94 cm^{-1} (P18) (the nutation damping at the frequency of the P18 CO_2 laser line in SF_6 is similar to that at the P20 line). We drew a similar conclusion about the type of transition that is resonant with the P16 line in SF_6 in our study of the photon echo in that gas (see Section 7). Figure 23c shows an oscilloscope trace of nutation in BCl_3 at the frequency of the P12 line.

Therefore, an experimental study of nutation in the molecular gases SF_6 and BCl_3 has made it possible to determine such characteristics of these gases as their dipole moment, absorption cross section, and relative number of particles involved in absorption. In addition, it was possible to establish that degeneracy has a substantial effect on nutation and to identify the type of transition responsible for nutation in these gases.

CONCLUSION

This article has dealt with a study of the coherent interaction of CO_2 laser light with the molecular gases BCl_3 and SF_6 and was intended to determine such characteristics of these gases as the dipole moments and types of the absorbing transitions, the longitudinal and transverse relaxation times T_2 and T_1, the absorption coefficients, and the fraction of active particles interacting with the radiation. Knowledge of these parameters is required for further exploitation of these gases in quantum electronics.

Here we have obtained the following basic results:

1. Self-induced transparency in BCl_3 has been studied and it has been found that degeneracy has an important influence on it. This allowed us to (a) isolate the range of pulse lengths and intensities whose behavior is similar to that of a π-pulse in a nondegenerate sys-

tem and to evaluate the matrix element of the dipole moment of the transition at 951.16 cm^{-1}; (b) evaluate the time T_2 in this gas from the way the shape of the pulse propagating in the medium changes with pressure.

2. The photon echo has been studied in BCl_3 and SF_6. This made it possible to (a) measure the time T_2 in these gases and evaluate the cross section for irreversible collisional dephasing from the variation of the photon echo intensity with the time separation between the exciting pulses; (b) find that a Q-branch transition with large J is responsible for absorption of the CO_2 laser P16 line in SF_6 based on the dependence of the photon echo polarization direction on the angle between the polarization directions of the exciting pulses; (c) study the concentration dependence of the photon echo; find the quadratic dependence of the photon echo intensity on the particle concentration for concentrations $n \leqslant 10^{10}$ cm^{-3}; find the deviation from quadratic of this concentration dependence at higher pressures; and estimate the radiative lifetime of the superradiant state.

3. The longitudinal relaxation time T_1 in SF_6 and BCl_3 has been measured.

4. The nutation effect in BCl_3 and SF_6 was studied and it was found that degeneracy and inhomogeneous broadening of the transition have a substantial influence on it. This made it possible to (a) measure the dipole moments of the transitions responsible for absorption of the P16, P18, and P20 CO_2 laser lines in SF_6 and of the P12 line in BCl_3; (b) having measured the absorption coefficients of SF_6 and BCl_3, to determine the fraction q_0 of the total number of particles within the Doppler profile of the absorption transitions, the fraction q of the total number of particles which takes part in absorption, and the homogeneous absorption cross section of these transitions (see the table).

The author is deeply grateful to his research adviser, Prof. N. V. Karlov, for proposing this topic, for valuable advice during the experiments and during the discussions of the results, and for constant encouragement. The author thanks Academician A. M. Prokhorov for his interest in this work, and Yu. B. Konev, O. M. Stel'makh, and I. V. Evseev for fruitful discussions and valuable comments.

LITERATURE CITED

1. F. Bloch, Phys. Rev., 70:460 (1946).
2. E. L. Hahn, Phys. Rev., 80:580 (1950).
3. I. D. Abella, N. A. Kurnit, and S. R. Hartmann, Phys. Rev. Lett., 13:567 (1964).
4. S. L. McCall and E. L. Hahn, Phys. Rev., 183:457 (1969).
5. C. K. N. Patel and R. E. Slusher, Phys. Rev. Lett., 19:1019 (1967).
6. C. K. N. Patel and R. E. Slusher, Phys. Rev. Lett., 20:1087 (1968).
7. G. B. Hocker and C. L. Tang, Phys. Rev. Lett., 21:591 (1968).
8. N. V. Karlov, G. P. Kuz'min, Yu. N. Petrov, and A. M. Prokhorov, Pis'ma Zh. Éksp. Teor. Fiz., 7:174 (1968).
9. I. D. Abella, N. A. Kurnit, and S. R. Hartmann, Phys. Rev., 141:391 (1966).
10. C. K. Rhodes, A. Shöke, and A. Javan, Phys. Rev. Lett., 21:1151 (1968); Phys. Rev., 184:1 (1969).
11. J. P. Gordon, C. H. Wang, C. K. N. Patel, R. E. Slusher, and W. J. Tominson, Phys. Rev., 179:294 (1969).
12. A. I. Alekseev and I. V. Evseev, Zh. Éksp. Teor. Fiz., 56:2118 (1969); Pis'ma Zh. Éksp. Teor. Fiz., 10:105 (1969); Izv. Akad. Nauk SSSR, Ser. Fiz., 37:2129 (1973).
13. R. H. Dicke, Phys. Rev., 93:99 (1954).
14. H. C. Torrey, Phys. Rev., 76:1059 (1949).
15. C. L. Tang and H. Startz, Appl. Phys. Lett., 10:145 (1967).
16. S. S. Alimpiev and N. V. Karlov, Zh. Éksp. Teor. Fiz., 61:1778 (1971).

17. N. V. Karlov, G. P. Kuz'min, A. M. Prokhorov, and V. I. Shemyakin, Zh. Éksp. Teor. Fiz., 54:1318 (1969).
18. S. S. Alimpiev, Yu. B. Konev, G. P. Kuz'min, and R. P. Petrov, Radiotekh. Élektron., 15:2361 (1970).
19. S. S. Alimpiev, Yu. B. Konev, G. P. Kuz'min and R. P. Petrov, Pis'ma Zh. Éksp. Teor. Fiz., 9:377 (1969).
20. S. S. Alimpiev and N. V. Karlov, Zh. Éksp. Teor. Fiz., 63:482 (1972).
21. S. S. Alimpiev and N. V. Karlov, Kratk. Soobshch. Fiz., No. 1, p. 17 (1973).
22. P. Rabinowitz, R. Kelle, and J. T. LaTourette, Appl. Phys. Lett., 14:376 (1969); M. W. Goldberg and R. Yusek, Appl. Phys. Lett., 17: (1970).
23. A. M. Ponte Goncalves, A. Tallet, and R. Leferbe, Phys. Rev., 188:576 (1969).
24. A. Compaan and I. D. Abella, Phys. Rev. Lett., 27:23 (1971).
25. G. Herzberg, The Vibrational and Rotational Spectra of Multiatomic Molecules [Russian translation], IL, Moscow (1949), p. 531.
26. L. M. Sverdlov, M. A. Kovner, and E. P. Krainov, Vibrational Spectra of Multiatomic Molecules [in Russian], Nauka, Moscow (1970), pp. 487 and 500.
27. S. S. Alimpiev and N. V. Karlov, Zh. Éksp. Teor. Fiz., 66:2 (1974).
28. P. G. Zykov, A. S. Rasponin, and P. E. Suetin, Zh. Éksp. Teor. Fiz., 43:200 (1973).
29. S. S. Alimpiev and N. V. Karlov, Izv. Akad. Nauk SSSR, Ser. Fiz., 37(10) (1973).

THE PHOTON ECHO IN GASES AT HIGH EXCITING PULSE INTENSITIES

S. S. Alimpiev, N. V. Karlov, and B. G. Sartakov

The formation of photon echoes in gases at high exciting pulse intensities is examined theoretically, including the inhomogeneous broadening of absorption lines and spatial degeneracy of the molecular states. A theoretical dependence of the photon echo in gases on the intensity of the exciting pulses is obtained. The results are compared with published experimental data.

The photon echo is one of the nonlinear coherent interactions of a resonant light field with matter. Coherence of an interaction means that the characteristic interaction time of the field with matter is much less than the relaxation time of the polarization induced in the medium by the field. The photon echo is basically the optical analog of the spin echo of NMR which arises in an ensemble of nuclear spins located in a constant magnetic field when a weak, variable magnetic field at a resonant frequency is applied. The first papers in this field are those by Bloch [1] and Hahn [2].

It became possible to observe the photon echo effect after the development of sources of coherent monochromatic radiation (lasers) and this has been done by a number of authors [3-5]. A photon echo is produced when a medium is illuminated successively by two exciting pulses, separated by a time interval T, whose frequencies coincide with that of an absorbing transition of the medium. The pulse emitted by the medium at time 2T is known as a photon echo. Here the inhomogeneous broadening of the transition (due to the Doppler effect in a gaseous medium) is important.

The photon echo effect is described theoretically in the framework of the resonant interaction of light with a set of two-level systems whose frequencies are somewhat off that of the field (inhomogeneous broadening). Each two-level system describes either an atom in a solid or a molecule in a gas. The photon echo problem was solved in a series of papers by A. I. Alekseev and I. V. Evseev [6] in which they limited themselves to the case of small values of the angular momentum J, which is essentially the case of photon echo on the electronic transitions of atoms. In molecular gases the formation of a photon echo is affected by the fact that the bulk of the molecules are usually in states with large values of J and are strongly degenerate with respect to the magnetic quantum number M. Degeneracy strongly affects the process whereby a photon echo is formed in gases and must always be taken into account when describing the photon echo phenomenon.

The photon echo in gases including degeneracies has been examined theoretically by Gordon et al. [7], who made several simplifying assumptions which limit the discussion to the case of an optically thin medium. In addition, they have neglected inhomogeneous broadening at the moment the exciting pulses act, which amounts to the condition

$$\Delta\omega_D\tau \ll 1,$$

where $\Delta\omega_D$ is the Doppler width of the spectral line and τ is the duration of the exciting pulses. Since the opposite relation

$$\Delta\omega_D\tau \gg 1$$

holds in all experiments on the photon echo in gases done up to this time [4, 5], the results of [7] cannot generally be used to analyze the available experimental data.

Here we derive a theoretical dependence for the photon echo intensity in gases on the intensity of the exciting pulses without using the condition $\Delta\omega_D\tau \ll 1$ and compare it with the dependences obtained in [7] and in Alimpiev's experiments [8].

We shall examine the photon echo effect in a molecular gas for the case of an optically thin layer where the spatial variation of the intensity of the light field may be neglected. We assume that the pulses which excite the photon echo are plane, linearly polarized waves propagating along the z axis of a Cartesian laboratory coordinate system, that the angle between the polarization vectors of the pulses is zero, and that the envelopes of the light fields obey the conditions

$$\begin{gathered}\int_{-\infty}^{+\infty} \frac{\mu_0 \mathcal{E}_1(t)}{\hbar}\, dt = \frac{\pi}{2}(2m+1),\\ \int_{-\infty}^{+\infty} \frac{\mu_0 \mathcal{E}_2(t)}{\hbar}\, dt = \pi(2m+1) \quad (m = 0,\ 1,\ 2,\ \ldots),\end{gathered} \tag{1}$$

where $\mathcal{E}_1$ and $\mathcal{E}_2$ are the envelopes of the first and second exciting pulses, respectively, and μ_0 is the largest matrix element of the dipole moment of the transition that is resonant with the field.

One feature of the photon echo in such molecular gases as SF_6 and BCl_3 [5] is that at room temperatures the bulk of the molecules are in states with $J \sim 40$–50. These states are degnerate in the magnetic quantum number M with a multiplicity of $2J + 1$. The existence of a spatial degeneracy in M has a considerable effect on self-induced transparency [9] and certainly affects the formation of a photon echo in gases.

With spherical-top molecules (such as SF_6) photon echo formation may be affected by a K-degeneracy (that is, a degeneracy in the projection of the vector **J** along an axis rigidly attached to the molecule), if the K-splitting is small, as well as by the degeneracy in M. It is, however, known that even in a spherical-top molecule the K-degeneracy is removed [10]. Thus, in our discussion we shall ignore K-degeneracy, assuming it to be removed and that the distance between neighboring K-split sublevels is greater than the Doppler width of the line.

Our chosen condition of an optically thin layer corresponds to the low-pressure case. We shall suppose that the pressure is so small that the condition

$$\Delta\omega_p \ll \Delta\omega_D,$$

where $\Delta\omega_p$ is the pressure broadening, is satisfied. Let the frequency of the exciting pulses, ω_0, coincide with the center of one of the K-split lines. Then only those molecules whose frequency is close to ω_0 will participate in formation of a photon echo. The density of these molecules in ω space is described by the customary formula

$$n(\omega) = \frac{N_{\omega_0}}{\sqrt{\pi}\,\Delta\omega_D} \exp\left[-\left(\frac{\omega-\omega_0}{\Delta\omega_D}\right)^2\right], \tag{2}$$

where N_{ω_0} is the number of molecules per cubic centimeter of gas whose transition frequency is ω_0 without the Doppler shift.

The existence of a spatial degeneracy in M means that the dipole moment of a transition from the state with quantum number M when J is large will be given by

$\mu = \mu_0 M/J$ for Q-branch transitions and by

$\mu = \mu_0 (1 - M^2/J^2)^{1/2}$ for P- or R-branch transitions.

For large J the density of molecules in ω space may be assumed continuous, so

$$n(\omega, \mu) = \frac{N\omega_0}{\sqrt{\pi}\,\Delta\omega_D \mu_0} \exp\left[-\left(\frac{\omega-\omega_0}{\Delta\omega_D}\right)^2\right] \text{ for the Q-branch, and} \tag{3}$$

$$n(\omega, \mu) = \frac{\mu}{\mu_0\sqrt{\mu_0^2-\mu^2}} \frac{N_{\omega_0}}{\sqrt{\pi}\,\Delta\omega_D} \exp\left[-\left(\frac{\omega-\omega_0}{\Delta\omega_D}\right)^2\right] \text{ for the P- and R-branches,} \tag{4}$$

where μ varies between 0 and μ_0. Thus, a molecular gas may be replaced by a set of two-level systems whose density in ω, μ space is given by Eq. (3) for the Q-branch and by Eq. (4) for the P- and R-branches.

We now turn to the behavior of a two-level system with a transition frequency ω close to the frequency of the laser pulses used to excite it and with a transition dipole moment μ. We denote the pulse lengths by τ and the time between them by T. Let τ and T satisfy the conditions

$$\tau \ll T_1,\ T_2; \qquad T \ll T_1,\ T_2,$$

where T_1 and T_2 are the longitudinal and transverse relaxation times [1]. Then the behavior of the two-level system may be described with the aid of the ψ function (wave function) which obeys

$$i\hbar\frac{\partial\psi}{\partial t} = \hat{H}\psi. \tag{5}$$

We shall seek the solution $\psi(t)$ of the Schrödinger equation in the form

$$\psi(t) = a_1(t)\,\psi_1(t) + a_2(t)\,\psi_2(t),$$

where $\psi_1 = \begin{Vmatrix} 0 \\ 1 \end{Vmatrix}$ is the ψ function of the ground state and $\psi_2 = \begin{Vmatrix} 1 \\ 0 \end{Vmatrix} e^{i\omega t}$ is the ψ function of the upper state. For $\hat{H}$ in the dipole approximation we have

$$\hat{H} = \begin{Vmatrix} 0 & 0 \\ 0 & \hbar\omega \end{Vmatrix} - E\mu \begin{Vmatrix} 0 & 1 \\ 1 & 0 \end{Vmatrix}, \tag{6}$$

where E is the electric field strength.

Equation (5), together with (6), yields the following equations for a_1 and a_2:

$$i\hbar\frac{\partial a_1}{\partial t} = -E\mu a_2 e^{-i\omega t},$$
$$i\hbar\frac{\partial a_2}{\partial t} = -E\mu a_1 e^{i\omega t}. \tag{7}$$

Let us transform to the new variables

$$u = \mathrm{Re}\,[a_1 a_2^* \exp(i\Delta\omega t + ikz)],$$
$$v = \mathrm{Im}\,[a_1 a_2^* \exp(i\Delta\omega t + ikz)],$$
$$n = {}^1/_2(|a_1|^2 - |a_2|^2),$$

where $\Delta\omega = \omega - \omega_0$, $k = \omega_0/c$, and z is the coordinate of a molecule.

We then have for the polarization P of the medium,

$$P = \mu[(u+iv)\exp(i\omega_0 t - ikz) + \text{c.c.}].$$

We write the electric field strength in the form

$$E = {}^1/_2\,\mathcal{E}(\theta)\,e^{i\omega_0\theta} + \text{c.c.},$$

where $\theta = t - z/c$ and $\mathcal{E}$ is the envelope of the field. For simplicity we shall assume that $\mathcal{E}=\mathcal{E}^*$. Then for u, v, and n we obtain

$$\dot{u} = -\Delta\omega v,$$

$$\dot{v} = \Delta\omega u + \frac{\mathcal{E}\mu}{\hbar} n,$$

$$\dot{n} = -\frac{\mathcal{E}\mu}{\hbar} v,$$

where the dot denotes differentiation with respect to θ, on proceeding from Eq. (7) and leaving out the rapidly oscillating terms. These equations were first derived by Bloch [1].

We now introduce the vector **R** in u, v, n space with coordinates

$$\mathbf{R} = \begin{Vmatrix} u \\ v \\ n \end{Vmatrix}.$$

Its equation of motion has the form

$$\dot{\mathbf{R}} = \mathbf{\Omega} \times \mathbf{R}, \tag{8}$$

where $\mathbf{\Omega} = \begin{Vmatrix} \mathcal{E}\mu/\hbar \\ 0 \\ -\Delta\omega \end{Vmatrix}$. Equation (8), which describes the behavior of a two-level system in a field, was derived in [11].

We shall make the following approximation. Suppose the exciting pulses, of length τ, are rectangular and separated by a time interval T (Fig. 1). In accordance with Eq. (1), the amplitude of the field in the first pulse satisfies

$$\frac{\mu_0 \mathcal{E}_1}{\hbar}\tau = \frac{\pi}{2}(2m+1), \qquad m = 0, 1, 2, \ldots$$

At the initial time $\theta_0 = -T - 2\tau$, the system is in the ground state,

$$\mathbf{R}(\theta_0) = \frac{1}{2}\begin{Vmatrix} 0 \\ 0 \\ 1 \end{Vmatrix}.$$

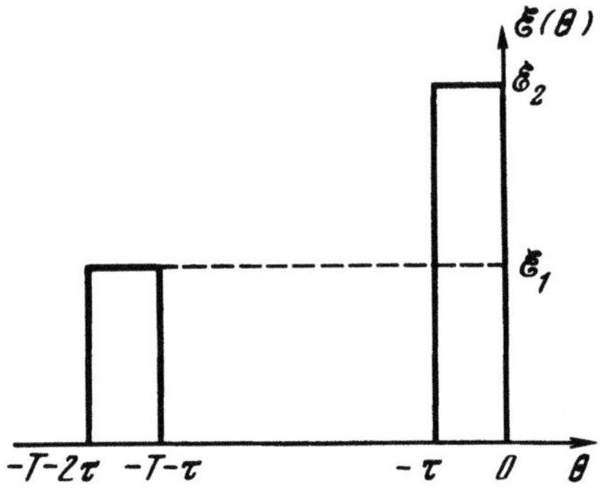

Fig. 1. A sketch of the pulses which excite the photon echo.

The time evolution of the vector **R** may be regarded as a rotation about the corresponding vectors $\mathbf{\Omega}$,

$$\mathbf{R}(\theta>0)=\hat{C}_3(t)\,\hat{C}_2\hat{C}_3(T)\,\hat{C}_1\mathbf{R}(\theta_0), \tag{9}$$

where $\hat{C}_{1,2,3}$ are the rotation matrices,

$$\hat{C}_{1,2}=\begin{Vmatrix} \cos^2\alpha_{1,2}+\sin^2\alpha_{1,2}\cos\varphi_{1,2} & \sin\alpha_{1,2}\sin\varphi_{1,2} & \cos\alpha_{1,2}\sin\alpha_{1,2}(1-\cos\varphi_{1,2}) \\ -\sin\alpha_{1,2}\sin\varphi_{1,2} & \cos\varphi_{1,2} & \cos\alpha_{1,2}\sin\varphi_{1,2} \\ \cos\alpha_{1,2}\sin\alpha_{1,2}(1-\cos\varphi_{1,2}) & -\cos\alpha_{1,2}\sin\varphi_{1,2} & \sin^2\alpha_{1,2}+\cos^2\alpha_{1,2}\cos\varphi_{1,2} \end{Vmatrix};$$

$$C_3(t)=\begin{Vmatrix} \cos\Delta\omega t & -\sin\Delta\omega t & 0 \\ \sin\Delta\omega t & \cos\Delta\omega t & 0 \\ 0 & 0 & 1 \end{Vmatrix}.$$

Here

$$\cos\alpha_{1,2}=\frac{\Omega_{1,2}}{(\Omega_{1,2}^2+\Delta\omega^2)^{1/2}};\qquad \sin\alpha_{1,2}=-\frac{\Delta\omega}{(\Omega_{1,2}^2+\Delta\omega^2)^{1/2}};$$

$$\varphi_{1,2}=\tau(\Omega_{1,2}^2+\Delta\omega^2)^{1/2};\qquad \Omega_{1,2}=\frac{\mathcal{E}_{1,2}\mu}{\hbar}.$$

To diagonalize the matrix $\hat{C}_3$ we go to the new variables

$$\xi=u+iv;\qquad \eta=u-iv;$$

then

$$C_{1,2}=\begin{Vmatrix} \frac{1}{2}(\cos^2\alpha_{1,2}+\sin^2\alpha_{1,2}\cos\varphi_{1,2}+ \\ +\cos\varphi_{1,2}-2i\sin\alpha_{1,2}\sin\alpha_{1,2}) & \cos^2\alpha_{1,2}\sin^2\left(\frac{\varphi_{1,2}}{2}\right) & \cos\alpha_{1,2}\sin\alpha_{1,2}(1-\cos\varphi_{1,2})+ \\ & & +i\cos\alpha_{1,2}\sin\varphi_{1,2} \\ \cos^2\alpha_{1,2}\sin^2\frac{\varphi_{1,2}}{2} & \frac{1}{2}\cos^2\alpha_{1,2}+\sin^2\alpha_{1,2}\cos\varphi_{1,2}+ & \cos\alpha_{1,2}\sin\alpha_{1,2}(1-\cos\varphi_{1,2})- \\ & +\cos\varphi_{1,2}+2i\sin\alpha_{1,2}\sin\varphi_{1,2} & -i\cos\alpha_{1,2}\sin\varphi_{1,2} \\ \frac{1}{2}[\cos\alpha_{1,2}\sin\alpha_{1,2}(1-\cos\varphi_{1,2})+ & \frac{1}{2}[\cos\alpha_{1,2}\sin\alpha_{1,2}(1-\cos\varphi_{1,2})- & \sin^2\alpha_{1,2}+\cos^2\alpha_{1,2}\cos\varphi_{1,2} \\ +i\cos\alpha_{1,2}\sin\varphi_{1,2}] & -i\cos\alpha_{1,2}\sin\varphi_{1,2}] & \end{Vmatrix}$$

and

$$C_3(t)=\begin{Vmatrix} e^{i\Delta\omega t} & 0 & 0 \\ 0 & e^{-i\Delta\omega t} & 0 \\ 0 & 0 & 1 \end{Vmatrix}.$$

By taking the product in Eq. (9) it is easy to separate the part of the expression for $\xi = u + iv$ which is proportional to $\exp[i\Delta\omega(\theta - T)]$ and, therefore, is responsible for the appearance of a photon echo at time $\theta = T$:

$$u+iv=e^{i\Delta\omega(\theta-T)}\frac{\Omega_2^2}{\Omega_2^2+\Delta\omega^2}\sin^2\left(\frac{\tau}{2}\sqrt{\Omega_2^2+\Delta\omega^2}\right)\left\{\frac{\Omega_1\Delta\omega}{\Omega_1^2+\Delta\omega^2}\left[1-\cos\left(\tau\sqrt{\Omega_1^2+\Delta\omega^2}\right)\right]-i\frac{\Omega_1}{\sqrt{\Omega_1^2+\Delta\omega^2}}\sin\left(\tau\sqrt{\Omega_1^2+\Delta\omega^2}\right)\right\}. \tag{10}$$

With this equation it is easy to find the polarization per unit volume, $P(\theta)$, remaining in the medium after the two exciting pulses have passed:

$$P(\theta)=\mathscr{P}(\theta)\,e^{i\omega_0\theta}+\text{c.c.}=\left[\int_0^{\mu_0}d\mu\int_{-\infty}^{+\infty}d\omega\,n(\omega,\mu)\,\mu\,(u+iv)\right]e^{i\omega_0\theta}+\text{c.c.}$$

This solution for the polarization per unit volume of the molecular gas must be related to the wave equation.

Let the gas lie between the planes z = 0 and z = l. For the field in the approximation of slowly varying amplitudes,

$$\left|\frac{\partial \mathcal{E}}{\partial \theta}\right| \ll |\omega_0 \mathcal{E}|, \qquad \left|\frac{\partial \mathcal{E}}{\partial t}\right| \ll |k\mathcal{E}|$$

the following equation is true:

$$\frac{\partial \mathcal{E}}{\partial z} = -2\pi i k \mathcal{P}(\theta), \tag{11}$$

where k is the wave vector and $\mathcal{P}(\theta)$ is the envelope of the polarization. Including the boundary conditions at the entry to the medium (z = 0),

$$\mathcal{E}(\theta, 0) = 0 \quad \text{for } \theta > 0,$$

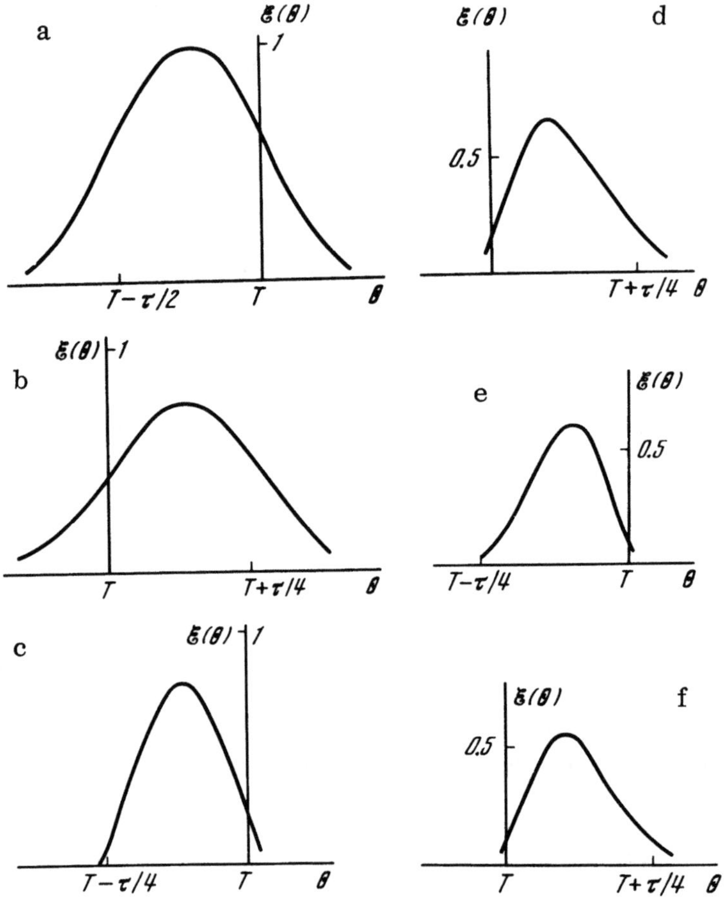

Fig. 2. Computer calculations of some photon echo pulse shapes produced by a Q-branch transition. a) m = 0; b–f) m = 1, 2, 3, 4, and 5, respectively. The photon echo pulse amplitude for m = 0 ($\pi/2$ and π exciting pulses) is taken to be unity.

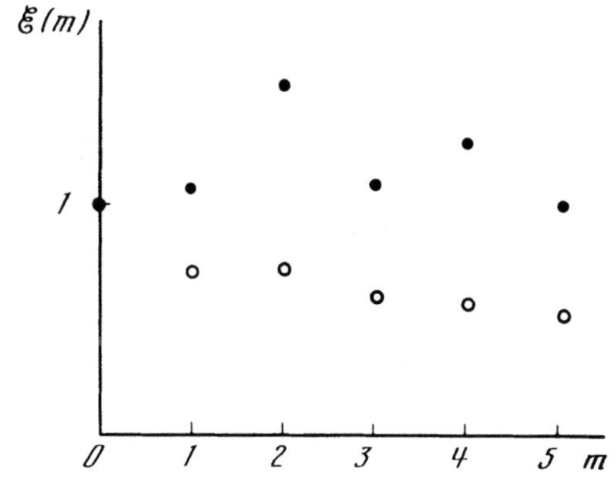

Fig. 3. Computed dependence of the photon echo pulse amplitude on m for Q-branch (circles) and P- and R-branch transitions (dots). The amplitude of the photon echo pulses is taken to be unity at m = 0.

we obtain from Eqs. (10) and (11) an expression for the photon echo pulse amplitude at the outlet from the medium:

$$\mathcal{E}(\theta > 0,\ l) = 8\pi k l \int_0^{\mu_0} d\mu \int_{-\infty}^{+\infty} d(\Delta\omega)\, n(\Delta\omega,\ \mu)\, \mu \frac{\Omega_2^2}{\Omega_2^2 + \Delta\omega^2} \sin^2\left(\frac{\tau}{2}\sqrt{\Omega_2^2 + \Delta\omega^2}\right) \times$$
$$\times \left\{ \frac{\Omega_1 \Delta\omega}{\Omega_1^2 + \Delta\omega^2} \left[1 - \cos\left(\tau\sqrt{\Omega_1^2 + \Delta\omega^2}\right)\right] \sin\left[\Delta\omega(\theta - T)\right] - \frac{\Omega_1}{\sqrt{\Omega_1^2 + \Delta\omega^2}} \sin\left(\tau\sqrt{\Omega_1^2 + \Delta\omega^2}\right) \cos\left[\Delta\omega(\theta - T)\right] \right\}. \quad (12)$$

This equation was used to compute the photon echo pulse shape on a computer. Figure 2 shows some calculations of the envelopes of photon echo pulses formed by a Q-branch transition with large J. The dimensionless variable m is related to the area under the first exciting pulse by

$$\frac{\mu_0 \mathcal{E}_1}{\hbar} \tau = \frac{\pi}{2}(2m + 1).$$

It was assumed that $\Delta\omega_D \tau = 8$.

From an analysis of the curves of Fig. 2 it is clear that as the area under the first pulse is increased (m = 0, 1, 2, 3, 4, 5) the duration of the photon echo pulse changes little, since when m is changed from 0 to 5, which corresponds to an increase in the intensity of the exciting pulses by two orders of magnitude, the duration of the photon echo pulse falls by only a factor of 2. This result is in agreement with the weak dependence of the duration of the photon echo in SF_6 and BCl_3 on the intensity of the exciting pulses observed experimentally by us. It is also clear from Fig. 2 that the time the photon echo appears differs somewhat from $\theta = T$, with the amount and sign of this difference determined by the intensity of the exciting pulses. A dependence of the time the photon echo appears on the area under the exciting pulses has been observed experimentally.

The calculated dependence of the amplitude of the envelope of the photon echo pulse on m for Q-, P-, and R-branch transitions is shown in Fig. 3. From this it is clear that the photon echo for Q-branch transitions is maximal when m = 0 ($\pi/2$ pulse). As m is increased the photon echo amplitude falls, but little. For P- and R-branch transitions it is found that the echo amplitude peaks at m = 2 ($5\pi/2$ pulse).

It should be noted that the way the curve of Fig. 3 (circles) behaves agrees with the experimentally observed dependence of the photon echo pulse intensity on the intensity of the exciting pulses in SF_6 at the frequency of the P16 CO_2 laser line [8].

Therefore, this analysis of the formation of a photon echo in a gas taking degeneracy into account and without limitations on the Doppler width of the line reflects the actual situation fairly completely and makes it possible to interpret many experimental observations.

Literature Cited

1. F. Bloch, Phys. Rev., 70:460 (1946).
2. E. L. Hahn, Phys. Rev., 80:580 (1950).
3. I. D. Abella, N. A. Kurnit, and S. R. Hartmann, Phys. Rev. Lett., 13:567 (1964).
4. C. K. N. Patel and R. E. Slusher, Phys. Rev. Lett., 20:1087 (1967).
5. S. S. Alimpiev and N. V. Karlov, Zh. Éksp. Teor. Fiz., 63:482 (1972).
6. A. I. Alekseev and I. V. Evseev, Zh. Éksp. Teor. Fiz., 56:2118 (1969); Pis'ma Zh. Éksp. Teor. Fiz., 10:105 (1969); Izv. Akad. Nauk SSSR, Ser. Fiz., 37:2129 (1973); Zh. Éksp. Teor. Fiz., 68:456 (1975).
7. J. P. Gordon, C. H. Wang, C. K. N. Patel, R. E. Slusher, and W. J. Tomlinson, Phys. Rev., 179:294 (1969).
8. S. S. Alimpiev, Trudy FIAN, 87:92 (present issue).
9. C. K. Rhodes, A. Shöke, and A. Javan, Phys. Rev., 184:1 (1969).
10. S. S. Alimpiev and N. V. Karlov, Zh. Éksp. Teor. Fiz., 66:3 (1974).
11. S. L. McCall and E. L. Hahn, Phys. Rev., 183:457 (1969).

THE RADIATIVE COHERENT INTERACTION OF LASER LIGHT WITH MATTER AND ITS APPLICATIONS

V. M. Akulin, S. S. Alimpiev, N. V. Karlov, and L. A. Shelepin

The resonant action of laser radiation on matter is examined. It is shown that under certain conditions when the levels are sufficiently dense due to their being broadened in the laser field a purely radiative stepwise excitation (dissociation or ionization) mechanism with a step size equal to the laser quantum may operate. A diffusion approximation appropriate for describing coherent effects is developed, and the population dynamics of an N-level oscillator is studied. The results are used to describe "prompt" dissociation and ionization of molecules and atoms. The role of collisions is briefly considered. Some applied possibilities for this mechanism including, in particular, laser isotope separation, are discussed.

INTRODUCTION

The resonant interaction of laser light with matter has been studied intensively in recent years. This problem is of importance in the stimulation of chemical reactions, the creation of strongly nonequilibrium states of matter, and the separation of isotopes. Two basic mechanisms have been considered for these interactions. The first of these is the absorption of a quantum by a specific transition which is resonant (the frequency of the transition being equal to the frequency of the electromagnetic wave), or multiphoton absorption, in which several quanta are required for the transition (between two levels or in the continuous spectrum). Of the papers on these problems, [1-5] are noteworthy. The second mechanism is collisional, including both resonant absorption of radiant energy in collisions and the redistribution of energy over the degrees of freedom and energy levels by collisions (see [5-12]).

In addition, in studies of resonant absorption of laser light by transitions between the ground and first excited states of one of the vibrational modes of multiatomic molecules there is evidence that the upper vibrational levels are radiatively excited; that is, the radiation is absorbed in a series of transitions.

The nonthermal interaction of powerful laser light with a molecular gas was first discovered in [13]. Stationary illumination of a gas whose molecular vibrational spectrum coincides with the frequency of the laser light led to a split between the vibrational temperature and the translational temperature and to dissociation of the molecules. Collisions played an important role in this process.

Somewhat later a number of authors [14, 15] observed and studied the so-called prompt (instantaneous) dissociation of molecules which takes place during pulsed excitation of resonant gases over a time much shorter than the time between gas kinetic collisions and is therefore a purely radiative process. Collisionless (prompt) dissociation is characterized by the existence of an intensity threshold (10^8-10^9 W/cm^2) as well as a threshold with respect to the total energy throughput.

Prompt dissociation has been used to produce an isotopically selective reaction between molecules of BCl_3 and oxygen [16]. Nevertheless, the mechanism for prompt dissociation is still unknown.

In the following we shall analyze a coherent radiative mechanism for the interaction of radiation with matter including transitions over a sequence of almost equidistant levels. This mechanism is possible due to the "broadening" of the levels in the strong field of the laser radiation. If the levels are sufficiently close, a multilevel resonant system or even a single quasicontinous background may be formed. Including this mechanism makes it possible to explain a wide range of phenomena which occur during the interaction of laser radiation with a gas of multiatomic molecules, including prompt dissociation. The basic features of this mechanism as applied to the excitation of vibrational levels of multiatomic molecules are examined in Section 1. The related kinetic problems are analyzed in Sections 2-4. In Section 5 the feasibility of prompt ionization and its use in isotope separation are discussed, and in Section 6 the prospects for realization of a coherent mechanism in atoms and the effect of collisions are discussed qualitatively.

1. Radiative Excitation and Dissociation of Multiatomic Molecules

We now turn to a concrete discussion of the effect of intense resonance on multiatomic molecules.

Radiation which is resonant for the first vibrational level in diatomic molecules will be nonresonant for higher vibrational levels due to the anharmonicity of the vibrations. In multiatomic molecules the radiation yield from a resonance when higher vibrational levels are excited may be compensated, in principle, by the existence of a high density of energy levels. This circumstance was pointed out by Isenor et al. [14]. In fact, the total number of vibrational levels (including combination levels) corresponding to the excitation of s vibrational quanta for l vibrational modes is given by the binomial coefficient C^s_{s+l+1}. At first this number increases rapidly with s (that is, with increasing vibrational excitation of the molecule upward on the energy scale). At $s \sim l$ the growth slows down. Each vibrational level has a rotational structure also containing a large number of levels. In the case of the BCl_3 molecule [which has two nondegenerate and two doubly degenerate vibrations ($l = 6$) and the rotational constants $B_1 = 0.1$ cm^{-1} and $B_2 = 0.05$ cm^{-1} of a symmetric top, if we include the Coriolis splitting in the excitation energy, which corresponds to the absorption of 1 or 2 laser light quanta (1-$2 \times h\nu_3$)] the distance between the rotational sublevels is 10^{-2} to $5 \cdot 10^{-2}$ cm^{-1}. Thus, for the lower vibrational levels the distance between the rotational sublevels is still rather large. In addition, for energies of order 3-$4 \times h\nu_3$ the energy between the levels is less due to the increased density of the levels and becomes comparable to the homogeneous width of a rotational level (10^{-3} cm^{-1}) at pressures of the order of several Torr at room temperature. The first vibrational levels can be excited by nonresonant absorption in a strong field which leads to "broadening" of the levels. A similar mechanism for diatomic molecules was considered in [17] and for an N-level anharmonic oscillator in [18]. After overcoming the first few (two or three) vibrational levels the absorption becomes resonant in view of the increased density of levels but finally leads to instantaneous dissociation. It should be noted that transitions which are resonant with the field are generally forbidden in the adiabatic approximation. It might be expected that the dipole moment of such transitions would be about two orders of magnitude less than the dipole moment of an allowed transition.*

---

* If some degenerate mode is excited whose spectrum also contains a large number of states in high levels, then in principle allowed transitions may be realized all the way up to the dissociation limit.

In order to find the condition for effective radiative dissociation we shall consider the time evolution of the density matrix ρ of a system of BCl_3 molecules being acted on by a laser field. From the equation of motion of the density matrix,

$$i\hbar\dot{\rho}=[\hat{H},\ \rho],\qquad \hat{H}=\hat{H}_0+\hat{H}_{\text{int}}, \tag{1}$$

with the initial condition

$$\rho(-\infty)=\exp\left\{-\frac{\hat{H}_0}{kT}\right\}$$

for the S-matrix of the transformation from the interaction representation to the Heisenberg representation, in the interaction representation we have the equation

$$i\hbar\dot{S}=\hat{H}_{\text{int}}S, \tag{2}$$

where $\hat{H}_0$ is the unperturbed Hamiltonian and $\hat{H}_{\text{int}}$ is the Hamiltonian of the interaction of a molecule with the field E. For the field $E=\mathcal{E}(t)\cos(\omega t+\varphi)$ we have

$$\hat{H}_{\text{int}\,mn}=\mathcal{E}\,[\hat{d}^{+}_{mn}e^{i(\Delta\omega_{mn}t+\varphi)}+\text{c.c.}], \tag{3}$$

where $\Delta\omega_{mn}=\omega_{mn}-\omega$ is the frequency difference between the field and transition frequencies due, for example, to anharmonicity; $\hat{d}^{+}$ and $\hat{d}^{-}\equiv\hat{d}^{+*}$ are the above-diagonal and below-diagonal components of the dipole moment operator, respectively; and m and n are the energy level indices.

The solution of Eq. (2) may be written in the form

$$S=T\exp\left\{-\frac{i}{\hbar}\int\limits_{-\infty}^{t}\hat{H}_{\text{int}}dt\right\}, \tag{4}$$

where T is the chronological ordering operator.

It is difficult in general to calculate the S matrix using Eq. (4); thus, it is reasonable to examine two limiting cases of Eq. (4) which give the characteristic parameters for effective penetration of the lower (I) level and population of the upper (II) levels.

I. Large frequency differences, $\Delta\omega \gg 1/\tau$ (τ is the pulse length). Then the integral in Eq. (4) transforms to

$$\int\limits_{-\infty}^{t}H_{\text{int}\,mn}dt=\mathcal{E}\left[\frac{d^{+}_{mn}}{\Delta\omega_{mn}}e^{i(\Delta\omega_{mn}t+\varphi)}+\text{c. c.}\right]. \tag{5}$$

In the random-phase approximation this operator commutes with itself at any time; thus, the sign of the T ordering in Eq. (4) may be omitted. Then

$$S=\exp\left\{\frac{-i\mathcal{E}(t)}{\hbar}\left[\frac{d^{+}}{\Delta\omega}e^{i(\Delta\omega t+\varphi)}+\text{c. c.}\right]\right\}. \tag{6}$$

The diagonal elements of the density matrix have the form

$$\rho_{nn}\sim\sum_{q}S_{nq}S^{-1}_{qn}, \tag{7}$$

where q is a rotational sublevel of the ground vibrational state (we assume that in the absence of a field only the vibrational ground state is populated). We now expand the right side in a series and insert $\langle n|$ and $|q\rangle$. Then the first nonzero term in the series will be the term with $(d^+)^n$, or, in other words, the one which is proportional to $\mathcal{E}^{2n}$. Then

$$\rho_{nn} \sim \sum_q \langle n|\left(\frac{d^+}{\Delta\omega}\right)^n|q\rangle\langle q|\left(\frac{d^-}{\Delta\omega}\right)^n|n\rangle\frac{\mathcal{E}^{2n}}{\hbar^{2n}} + \ldots .$$

From this it follows that ρ_{nn} increases considerably when a threshold laser intensity,

$$I_{thr} \simeq \frac{c\hbar^2}{4\pi}\left(\sum_q \langle n|\left(\frac{d^+}{\Delta\omega}\right)^n|q\rangle\langle q|\left(\frac{d^-}{\Delta\omega}\right)^n|n\rangle\right)^{-1/n}, \tag{8}$$

is attained. Thus, from Eqs. (6) and (7) it follows directly that when a vibrational level whose energy is equal to the energy of n laser quanta is excited, the excitation efficiency is proportional to I^n, the effect has a threshold character, and the threshold intensity is given by Eq. (8).

II. Small frequency differences, $\Delta\omega \ll 1/\tau$. In this case the integral of Eq. (4) has the form

$$\int_{-\infty}^{t} \hat{H}_{int}dt = \hat{d}\int_{-\infty}^{t} \mathcal{E}(t)\,dt \tag{9}$$

and, therefore,

$$S_{mn} = \exp\left\{\frac{-i}{\hbar}\hat{d}\int_{-\infty}^{t}\mathcal{E}(t)\,dt\right\}_{mn}. \tag{10}$$

Equation (10) is analogous to the expression obtained during the analysis of coherent effects such as the photon echo in a two-level system. The basic difference is that the molecules are moved upward due to the radiation more rapidly than the relaxation rate. In view of this it is possible to neglect relaxation of the nondiagonal matrix elements ρ_{mn} at sufficiently high intensities. As will be clear from the estimates of τ_{min} in BCl_3 (T_2 = 27 nsec · Torr [19]) given below, relaxation may be neglected up to pressures of the order of several Torr.

We now expand Eq. (10) in a series analogous to that for Eq. (6):

$$\rho_{nni} \sim \sum_q \frac{1}{n!}\langle n|(\hat{d}^+)^n|q\rangle\langle q|(\hat{d}^-)^n|n\rangle\left(\frac{1}{\hbar}\int_{-\infty}^{t}\mathcal{E}(t)\,dt\right)^{2n} + \frac{O(1)}{(n+2)!}\left(\frac{\hat{d}}{\hbar}\int_{-\infty}^{t}\mathcal{E}(t)\,dt\right)^{2n+2}. \tag{11}$$

From this it is clear that for excitation of the upper vibrational levels the threshold in the integral of the pulse envelope rather than the intensity threshold becomes important, i.e., the important quantity is

$$\int_{-\infty}^{+\infty}\mathcal{E}(t)\,dt \sim \frac{\hbar n}{d_{forb}}.$$

For a given value of the intensity this condition essentially places a limitation on the minimum duration of the exciting pulse.

This analysis therefore shows that, in fact, in the case of multiatomic molecules the excitation of the upper vibrational levels all the way to dissociation is possible by a purely radiative mechanism which takes into account both the "broadening" of the levels due to the strong field and the high density of energy levels of the molecule. In a real situation the actual excitation channel would be some sort of synthesis of the limiting cases of large and small differences between the field and transition frequencies discussed here. During excitation of the several lowest-lying vibrational levels, the first limiting case of large frequency differences, but allowed transitions, is realized and this places a limitation on the intensity of the exciting pulse. During population of the higher vibrational states, all the way up to the dissociation limit, the second limiting case of forbidden resonant transitions is realized and this places a limitation on the minimum duration of the exciting pulses.

We now estimate the threshold intensity for excitation of the first two or three vibrational levels proceeding from Eq. (8),

$$I_{thr} \simeq \frac{c}{4\pi}\left(\frac{\hbar\Delta\omega}{d}\right)^2.$$

For the specific case of the BCl_3 molecule, which has a dipole moment for a vibrational transition of $d \sim 3 \cdot 10^{-19}$ cgse [19] and an anharmonicity $\Delta\omega \sim 3\ cm^{-1}$, the threshold intensity is $10^8\ W/cm^2$, which agrees with the known experimental data [14, 15].

The existence of a second mechanism in the excitation cascade makes the degree of dissociation depend on the pulse duration. This has also been observed experimentally [14]. As noted above, forbidden transitions take place in the second limiting case. The dipole moments of forbidden transitions calculated in a nonadiabatic approximation for the HD molecule [20] were found to have a similar value for all rotationally forbidden lines given by $d_{forb} \sim d(m_e/M)^{1/2}$, where m_e is the mass of an electron, M is the reduced nuclear mass, and d is the dipole moment of an allowed vibrational transition. Since these values of d_{forb} are caused by destruction of the symmetry of the rotational ψ functions, for multiatomic molecules with a lower degree of symmetry this would be an underestimate.

An estimate of the minimum duration of the exciting pulses according to Eq. (12), if we start with a threshold intensity of $10^8\ W/cm^2$ and a forbidden transition dipole moment of $d_{forb} \sim 10^{-2}\ d$, yields $\tau_{min} \sim 10^{-8}$-10^{-9} sec. We note that this value of τ_{min} is less than both the irradiating pulse duration and the time T_2.

This analysis has demonstrated the characteristic features of radiative dissociation. First, all the specifics of the molecular bond are lost, all the levels are "the same," and they form a single quasicontinuous background. Second, here the "bottleneck" is at the lower vibrational levels where the distance between the rotational sublevels is greatest; hence, there is a selectivity for molecules which exist under different conditions or contain different isotopes. Third, radiative dissociation may be controlled not only by changing the frequency, but also by changing the power and duration of the exciting pulse.

We note that dissociation may occur at low pressures where the broadening of the levels required for effective realization of the second limiting case is achieved by means of the strong field present in accordance with the first limiting case of our analysis, rather than by means of collisions.

2. Kinetic Analysis. An N-level Oscillator in a Resonant Field

The qualitative features of the radiative coherent mechanism are already clear from the previous analysis. However, for a quantitative analysis including a determination of the distribution function over the levels and its dependence (which also yields the dissociation rate), a detailed calculation of radiative processes is needed. We shall examine the kinetics of resonant coherent excitation of high vibrational levels of multiatomic molecules in a powerful laser field. For this purpose in the following we shall develop a diffusion approach and illustrate the procedure by constructing an S-matrix which describes the time evolution of the density matrix of an arbitray multiatomic molecule acted on by a laser field. The equation for the S-matrix has the form

$$i\dot{S} = \hat{H}_{\text{int}} S, \tag{12}$$

where $\hat{H}_{\text{int}} = \hat{d}(t)E$ is the Hamiltonian of the interaction of a light field of amplitude E with a a molecule, $\hbar = 1$, and $\hat{d}(t)$ is the dipole moment operator in the interaction representation. The solution of this equation is most conveniently written in the form

$$S_{mn}(t) = \delta_{mn} + (-i)\int_{-\infty}^{t} \hat{H}_{\text{int}\,mn}(t_1)\,dt_1 + (-i)^2 \int_{-\infty}^{t} \hat{H}_{\text{int}\,m,j}(t_1) \times$$
$$\times \int_{-\infty}^{t_1} \hat{H}_{\text{int}\,j,n}(t_2)\,dt_2 dt_1 + \ldots + (-i)^l \int_{-\infty}^{t} \hat{H}_{\text{int}\,m,j_1}(t_1) \ldots \int_{-\infty}^{t_{l-1}} \hat{H}_{\text{int}\,j_{l-1},n}(t_l)\,dt_1 \ldots dt_l. \tag{13}$$

The diffusion approach is also useful for describing the population dynamics of multiatomic molecules. Since in this model cascade dipole transitions are Markov processes, it is possible to expand the density matrix in powers of 1/N (where N is the total number of levels) when the number of levels is large compared to unity.

In this case the discrete indices of the energy levels are replaced by the continuous variables x = n/N and y = m/N. This makes it possible to reduce the problem to solving Kolmogorov's equation (see, for example, [21]), which in our case has the form

$$i\dot{\rho} = E\Big\{[d^+(x) + d^-(x)]\cosh\frac{\partial}{N\partial x} - [d^+(y) + d^-(y)]\cosh\frac{\partial}{N\partial y} +$$
$$+ [d^+(x) - d^-(x)]\sinh\frac{\partial}{N\partial x} - [d^+(y) - d^-(y)]\sinh\frac{\partial}{N\partial y}\Big\}\rho. \tag{14}$$

Here d^+ and d^- are the above- and below-diagonal components of the dipole moment operator. In this case the matrix ρ may be factored $\rho(x, y, t) = X(x, t)Y(y, t)$ and the problem reduced to solution of these equations:

$$i\dot{X} = E\Big[(d^+ + d^-)\cosh\frac{\partial}{N\partial x} + (d^+ - d^-)\sinh\frac{\partial}{N\partial x}\Big]X,$$
$$i\dot{Y} = -E\Big[(d^+ + d^-)\cosh\frac{\partial}{N\partial y} + (d^+ - d^-)\sinh\frac{\partial}{N\partial y}\Big]Y. \tag{15}$$

The problem of an N-level harmonic oscillator is of interest for a qualitative description of the population dynamics.

We shall consider the solution of Eqs. (15) for an N-level oscillator. For $d^+ \equiv d^- \equiv d$ these equations may be written in the form

$$iX'_\tau = \cosh\frac{\partial}{N\partial x}X, \quad iY'_\tau = \cosh\frac{\partial}{N\partial y}Y, \tag{16}$$

where we have transformed from the variable t to a new variable $\tau = 2d\int \mathcal{E}(t)\,dt$. To solve Eqs. (16) we shall apply the boundary conditions

$$\alpha X'_x + \beta X = 0|_{x=1}, \quad X'_x = 0|_{x=0},$$
$$\alpha Y'_y + \beta Y = 0|_{y=1}, \quad Y'_y = 0|_{y=0}.$$

The case $\beta \ll \alpha$ corresponds to small loss of particles from the system from the upper limiting

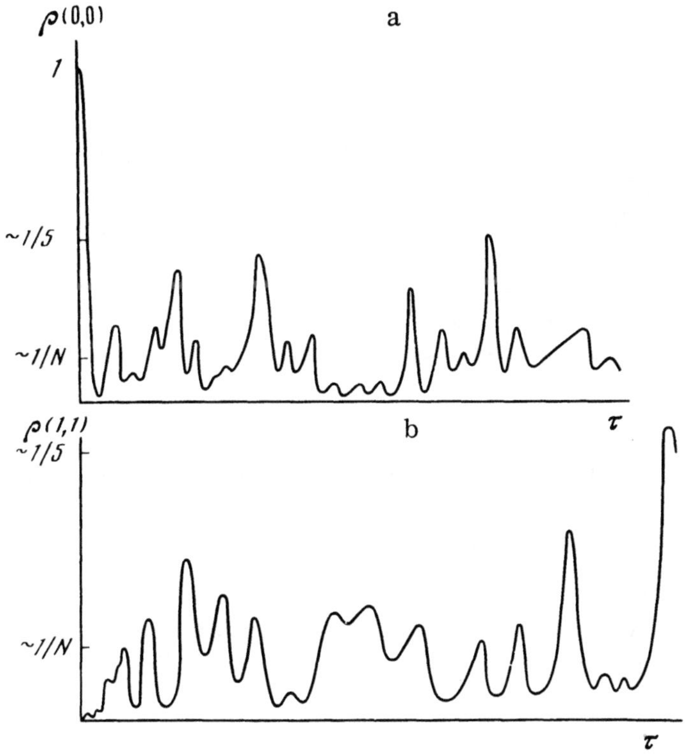

Fig. 1. The typical forms of the "fluctuating" populations of the lower (a) and upper (b) levels of a truncated harmonic oscillator.

level. The initial condition $\rho(x, y, \tau = 0) = \delta(x-0, y-0)$ refers to finding the system in the ground state in the absence of a field ($t = -\infty$). Under these conditions the solutions of Eqs. (16) can be expressed in terms of the Chebyshev polynomials $U_n(x)$:

$$\rho(x, y, \tau) = \rho_{n, m}(\tau) = \frac{1}{N} \sum_{p, q=1}^{N} e^{i(z_p - z_q)} U_n(z_p)\, U_m(z_q); \quad U_N(z_{p, q}) = 0; \tag{17}$$

z_p, z_q are roots of U(z).

The value of $\rho(x, x, \tau)$ determines the population of the level n = Nx at time τ. For a qualitative analysis of the populations of the vibrational levels of the molecule, Eq. (17) was evaluated on a computer. The results are shown in Fig. 1.

It is clear that $\rho(1, 1)$ is a function of the parameter τ localized in the neighborhood of the mean value $\overline{\rho(1, 1)} \sim 1/N$ (where N is the total number of levels) with rare sharp spikes which exceed the mean value by several times (these are the "fluctuating" populations).

The quantity $\rho(0, 0)$ falls rapidly at first and attains the "fluctuating" population form near the value 1/N.

3. The Population Dynamics of a Multilevel System in the Presence of a Single Nonresonant Level

In this section we shall examine the population dynamics of a multilevel system with the following model: We assume that the lower level is displaced from the field frequency by an amount $\Delta\omega$ and that the remaining levels are in resonance. The expressions which will be obtained below qualitatively describe the case in which the number of nonresonant transitions is

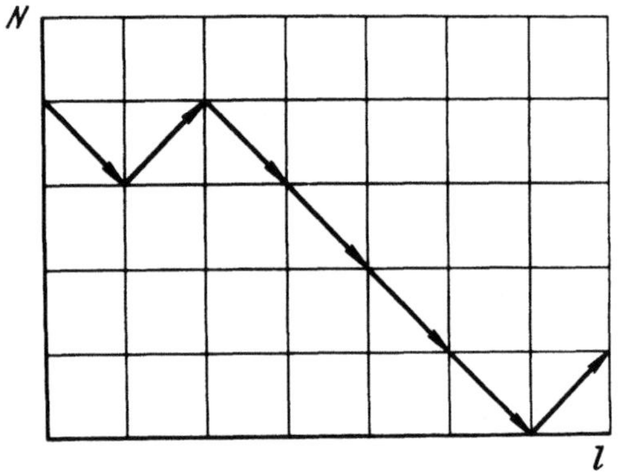

Fig. 2. The trajectory corresponds to the term in the series (18) with m = 4; n = 1; $\Delta i_1 = -1$; $\Delta i_2 = +1$; $\Delta i_3 = -1$; $\Delta i_4 = -1$; $\Delta i_5 = -1$; $\Delta i_6 = +1$.

small compared to the number of resonant ones. However, the effect of the number of nonresonant transitions is unimportant in most estimates and this procedure generalizes recurrently to an arbitrary number of displaced levels.

To describe a system with one displaced transition we shall use Eq. (12) and its solution (13) in the case where the field is strictly harmonic with amplitude E = const. The l-th order term in the series (13), which stands in the m-th column and the n-th row, may be written in the form of a sum,

$$\sum E^l \int_{-\infty}^{t} d_{n,\, n+i_1}(t_1) \int_{-\infty}^{t_1} d_{n+i_1,\, n+i_2}(t_2) \ldots \int_{-\infty}^{t_{l-1}} d_{n+i_l,\, m}(t_l)\, dt_1 \ldots dt_l, \tag{18}$$

where the sum is taken over various sets of indices $\{i_1; i_2; \ldots; i_l\}$ such that $i_{k+1} = i_k \pm 1$. For the summation we shall use a graphical technique. Each term of such a sum may be represented graphically in the form of a trajectory on a grid with N cells in the vertical and l along the horizontal (see Fig. 2). An arrow directed downward and to the right corresponds to $\Delta i_k \stackrel{\Delta}{=} i_k - i_{k-1} = +1$, and one directed upward and to the right, to $\Delta i_k = -1$. The direction of the arrows agrees with the order of integration.

The series (13) may be represented as a sum of all possible trajectories of the type shown in Fig. 2. We shall denote those portions of the trajectory which join the zeroth and first level by a broken curve (upside-down V) and the trajectories over the upper (N − 1) levels by a smooth curve. A bold upside-down V is used to denote the sum of all trajectories over the upper levels, and a wavy line to denote the sum of all the upside-down V's in succession. The diagrams for S_{m0} are shown in Fig. 3. Taking the graphical sum, we obtain the series shown in Fig. 4 for S_{m0}.

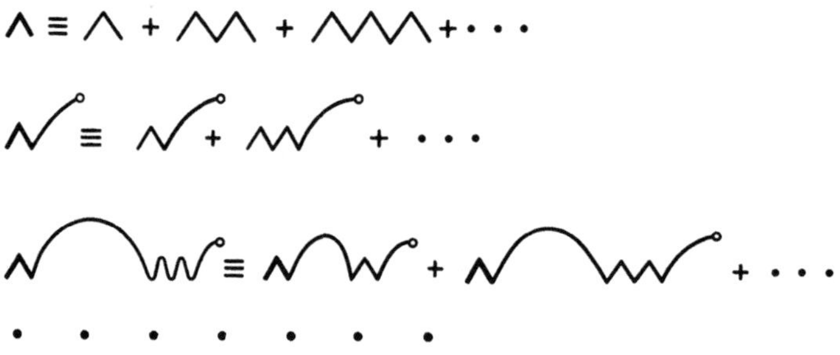

Fig. 3

Fig. 4

Using the operator Λ (bold upside-down V) in matrix form,

$$\Lambda = \int_\Omega \int_\Omega \left[\Delta\omega \begin{Vmatrix} 1 & 0 \\ 0 & 1 \end{Vmatrix} - \lambda_1 \begin{Vmatrix} 1 & 0 \\ 0 & 1 \end{Vmatrix} \right]^{-1} \left[\begin{Vmatrix} \Delta\omega & Ed \\ Ed & -\Delta\omega \end{Vmatrix} - \lambda_2 \begin{Vmatrix} 1 & 0 \\ 0 & 1 \end{Vmatrix} \right]^{-1} e^{i(\lambda_1+\lambda_2)t} d\lambda_1 d\lambda_2, \tag{19}$$

and using the formula for a function of an operator (see, for example, [22]),

$$f(\hat{A}) = \int_\Omega (z - \hat{A})^{-1} f(z)\, dz, \tag{20}$$

where $\hat{A}$ is an arbitrary nondegenerate linear operator and Ω is the profile embracing all the eigenvalues of the operator $\hat{A}$, we obtain

$$S_{m0}(-\infty,\ t) = \int_\Omega \int_\Omega (d_3 - \hat{d}_3)^{-1}_{11} (d_3' - \hat{d}_3)^{-1}_{1m} \left(1 - \frac{Ed_3 - \Delta\omega}{\Delta\omega \sqrt{\Delta\omega^2 + E^2 d^2}}\right)^{-1} \times$$
$$\times \exp\left[i\left(\sqrt{\Delta\omega^2 + E^2 d^2} - \Delta\omega\right)t\right] \left(1 - \frac{\Delta\omega^2}{\Delta\omega^2 + E^2 d^2 + E^2 d_3'^2}\right)^{-1} dd_3 dd_3', \tag{21}$$

where $\hat{d}_3$ is the matrix of the dipole moments of the resonant levels, d_3 and d_3' are variables, and d is the dipole moment of the first level displaced by $\Delta\omega$.

The other matrix elements S_{mn} with $n \geq 1$ no longer begin with an upside-down V but with the "tail," and their sum is thus:

$$S_{mn}(-\infty,\ t) = \int_\Omega (d_3 - \hat{d}_3)^{-1} \exp(id_3Et)\, dd_3 + \int_\Omega\int_\Omega\int_\Omega (d_3 - \hat{d}_3)^{-1}_{n1} (d_3' - \hat{d}_3)^{-1}_{11} (d_3'' - \hat{d}_3)^{-1}_{1m} \times$$
$$\times \frac{\exp(id_3t) \cdot dE^3 [d_3'/(d_3 - d_3'')] \exp(i\Delta\omega t)\, dd_3 dd_3' dd_3''}{(\Delta\omega + Ed)\left[(Ed_3 - \Delta\omega) Ed_3 + \Delta\omega\left(\sqrt{\Delta\omega^2 + d^2E^2} - \Delta\omega\right) d_3/(d_3' - d_3)\right]}. \tag{22}$$

Thus, the S-matrix has been found for the case of a strictly harmonic field E = const, which is turned on at $t = -\infty$. To find the populations of the levels when the field is turned on instantaneously it is necessary, evidently, to take

$$\hat{U}(t) = \begin{cases} \hat{\delta}, & t \leqslant 0, \\ \hat{S}^{-1}(-\infty,\ 0)\, \hat{S}(-\infty,\ t), & t > 0, \end{cases} \tag{23}$$

where $\hat{\delta}$ is the Kronecker delta, as an evolution operator.

The matrix $S^{-1}(-\infty, 0)$ is the inverse of $S(-\infty, 0)$ which is given by Eqs. (21)-(22). From an analysis of Eqs. (21) and (22) it follows that for $E > \Delta\omega/d$ all its elements are of order unity. Thus, the elements of the matrix U(t) are linear combinations of all the terms in one row of the S-matrix with coefficients $S_{n,m}$ of order unity, i.e.,

$$\rho_{nn} = \frac{1}{Z} \left| \int_\Omega \sum_{m=1}^{N} f_{nm}(d_3)\, e^{id_3Et} dd_3 + s_{n0+} e^{i\sqrt{\Delta\omega^2 + E^2d^2}} + s_{n0-} e^{-i\sqrt{\Delta\omega^2 + E^2 d^2}} \right|^2, \tag{23a}$$

where Z is a normalizing factor and f_{nm} is defined by $(n \geq 1)$

$$S_{mn}(-\infty,\ t) = \int_\Omega dd_3 f_{nm}(d_3)\, e^{-id_3Et} \tag{24}$$

[cf. Eq. (13)] and has simple poles at d_3 coinciding with the eigenvalues of the operator d_3. For $E \ll \Delta\omega/d$ the upper levels are populated with a low probability $\sim(Ed)^2/\Delta$, as is shown in [23], and may be directly derived from Eqs. (21)-(24).*

4. Analysis of the Populations

According to Eqs. (23) and (24), the basic feature of the population kinetics of an N-level multiatomic molecule is the eigenvalue spectrum of the dipole moment operator since this spectrum, to within a factor, agrees with the spectrum of frequencies ρ_{nn}, and, therefore, determines the time dependence. We shall examine a system in which the resonant transitions have the same dipole moment d_3 independent of the number of the level n (i.e., beginning with the first level it is a harmonic oscillator). We shall find the eigenvalue spectrum of the dipole moment operator for the resonant transitions in this case. Let D be the determinant of the $k \times k$ matrix in the upper-level corner of the matrix $\hat{d}_3 - d_3\hat{\delta}$. Then the following relationship is obvious:

$$D_k = -d_3 D_{k-1} - d_3^2 D_{k-2}, \tag{25}$$

Without restricting the generality we may set $d_3 = 1$ and get

$$\hat{d}_{3mn} = \delta_{n,\,m-1} + \delta_{n,\,m+1}.$$

Then, using the recurrence relation for the Chebyshev polynomials or a procedure analogous to that used in Eq. (14), we have

$$D_n = AU_n\left(\frac{d_3}{2}\right) + BT_n\left(\frac{d_3}{2}\right); \tag{26}$$

where A, B = const and the boundary conditions are

$$D_1 = d_3; \quad D_2 = d_3^2 - 1.$$

Solution of Eq. (26) yields the eigenvalue spectrum of the operator,

$$d_3 = 2\cos\frac{q\pi}{N+1}; \quad q = 1, 2, \ldots, N.$$

Since this spectrum determines the frequency spectrum of ρ_{nn} we may say that the lower boundary of the pairwise differences in the eigenvalues characterizes the minimum frequency in the spectrum of ρ_{nn}, and the upper boundary, the maximum frequency. Since ρ_{nn} is an oscillatory function [as can be seen directly from Eq. (24)], it is natural to assume that the maximum frequency defines the risetime of its (ρ_{nn}) time dependence, $\tau_r \sim (2\pi EdN^2)^{-1}$, and the least common multiple defines its repetition frequency, $\omega_n \gg 2\pi Ed$. It is easy to see that the spectrum ρ_{nn} obtained in this section is in good agreement with the spectrum $\rho(1, 1)$ obtained with the diffusion approach. The spectrum of ρ_{nn} has one additional frequency, $(\Delta\omega^2 + E^2d^2)^{1/2} - \Delta\omega$, compared with the spectrum of $\rho(1, 1)$.

The different dipole moments of the transitions between the levels are capable, generally speaking, of changing the spectrum; however, it may be said that this change would have little effect on the pattern of "fluctuating" populations.

*If $n \sim N \gg 1$, or $E \gg \Delta\omega/d$, then d_3Et may be replaced by $d_3\int Edt$ in Eq. (15).

The dissociation outflow may be taken into account by using a nonhermitian Hamiltonian. When there is a small outflow into dissociation the spectrum of $\hat{d}_3$ is slightly shifted in the complex plane so that Re d_3, Im $d_3 > 0$.

The numerical value of the outflow is determined by the population of the upper level and the probability of a transition into a dissociated state. For weak outflow it is possible to average the population of the upper level over the repetition period and set $\rho_{NN} \sim 1/N$.

5. Prompt Ionization and Laser-Electrical Isotope Separation

In the preceding we have examined the radiation coherent excitation of a system of vibrational–rotational levels of multiatomic molecules lying below the dissociation limit. However, between the dissociation and ionization limits there lies a special region which, if it is not taken into account, will leave the analysis of the coherent mechanism incomplete. This is also a region with a high density of levels (electronic, vibrational, rotational), but with special properties. Here the probability of the system's decaying due to rapid radiative transitions, predissociation, and other nonadiabatic decay processes, as well as dissociating electronic terms, becomes significant. This region contains a large number of electronic terms. As is known, the number of terms of a two-atom molecule made up of atoms with orbital and spin angular momenta (L_1S_1) and (L_2S_2) is given by

$$(2L_2+1)(L_1+1)(2S_1+1)$$

(here $L_1 < L_2$ and $S_1 < S_2$).

The exact structure of the terms of multiatomic molecules is poorly known. The terms of the four-atom molecule BCl_3 are classified according to the C_{3v} group and the molecule is made up of B and Cl atoms in the ground 2p states. The total number of electronic terms lying below the ionization limit is much greater than 100.

These terms are divided up into bonding and antibonding (or dissociating) terms. The first of these have a vibrational–rotational structure similar to the electronic ground state. A structure similar to that of the diatomic molecules N_2, Cl_2, and CO is typical of multiatomic molecules when a higher potential well lies within a lower one and forms a quasicontinuous chain of vibrational–rotational levels up to the ionization limit. If we consider only these terms, then the resulting pattern is mainly similar to that considered in Section 1. The differences are, first of all, that the dipole moments of the transitions between the electronic vibrational–rotational levels (as opposed to the vibrational–rotational transitions up to the dissociation limit) have been taken into account; and, secondly, that radiative transitions have a high probability and that nonadiabatic, including predissociation, transitions exist. This leads to the disintegration of the system. However, if the pulse duration is of order 10^{-9} sec, then this distinction is also unimportant. As a result, "prompt" (instantaneous) ionization takes place. This effect may be important for the chain of vibrational–rotational levels considered here. The fundamental factor affecting the kinetics and opening up another excitation and ionization channel is the existence of dissociating terms. In the absence of a field their lifetimes are of the order of 10^{-15} sec. Dissociating terms have a wide width (of order 1 eV) and a correspondingly small probability of radiative transitions (over these terms).

When a molecule enters a dissociating term that means it will disintegrate and leave the coherent excitation pattern. True, here we must include two kinds of corrections. First, the lifetime in a strong field may be greatly increased due to mixing of states. Second, as the fragments fly apart they conserve their energy (unless emission took place) until they collide with another fragment or molecule. At the time it collides, a molecule may go into a higher

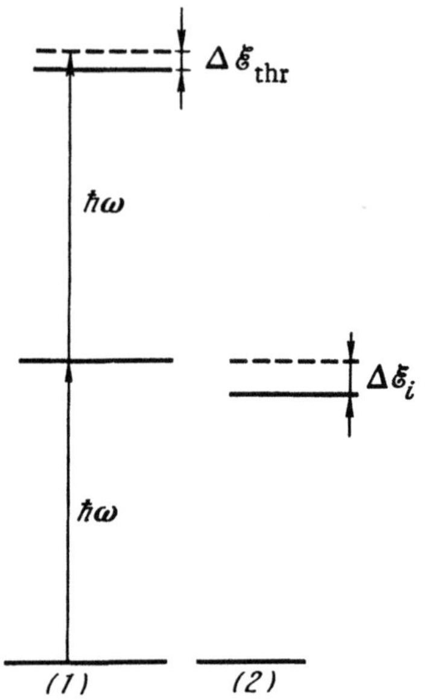

Fig. 5

term. Nevertheless, the existence of a second channel appears unavoidable. Clearly, we may speak of a relation between the prompt ionization of a molecule and its decay with subsequent ionization of the fragments.

A fundamental aspect of the purely radiative ionization mechanism to be considered later is the fact that the bottleneck for both ionization channels corresponds to transitions between lower vibrational levels; thus, there is a threshold selectivity for atoms containing different isotropes, and practically complete separation may be achieved in principle.

A possibility for laser-electrical isotope separation is based on the collisionless ionization effect. The products formed during prompt ionization are easily separated by an electric field. Then the ions formed in both channels belong to a single isotope. (The prompt ionization products are bonded chemically.)

The laser field intensity is limited above and below for effective isotope separation. The essence of these limitations is clear from Fig. 5, which is a schematic representation of the levels for isotopes (1) and (2). The field must be high enough that the broadening exceeds $\Delta\mathcal{E}_{thr}$ and at the same time not be so high that the broadening exceeds $\Delta\mathcal{E}_i$.

We note that in experiments on prompt dissociation [24] unusual products not associated with thermal dissociation or with breaking of a certain bond have been observed. Clearly, this is indirect evidence of the formation of products by dissociating terms and of the high extent to which molecules are excited in the region between the dissociation and ionization limits.

Prompt ionization has been observed experimentally (see [25]). When a molecular gas was irradiated by infrared laser light in a constant electric field for a time less than the V–V exchange time, an ion current was observed along the electric field. That is, charged heavy particles were detected in the experiment. However, it has not yet been possible to separate the two collisionless ionization channels.

We emphasize that both collisionless dissociation and collisionless ionization may be used as the basis of extremely effective isotope separation using the resonant interaction of a laser with a specific vibrational transition of a multiatomic molecule.

We note, however, that definite limitations result from the finite decay time τ_d and the need to satisfy the conditions on $\int \mathcal{E}(t)\,dt$ (see Section 1).

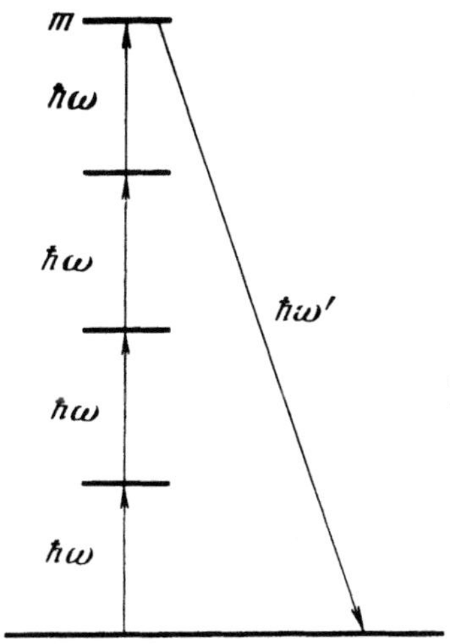

Fig. 6. A scheme for coherent frequency conversion.

If collisions take place, energy transfer to other isotopes is very efficient and the selectivity may be lost. Collisions may also have an effect after the end of the laser pulse. To exclude collisions a molecular flow may be used. At a pressure of about 1 Torr the time between collisions is about 10^{-6} sec and the appropriate dispersion velocity is easily achieved. Partial separation may be realized even if some of these limitations are not fulfilled. The efficiency of the laser-electrical (or laser-chemical) technique is determined by the concrete conditions in this case. We note that, in addition to isotope separation, some other important applications of this technique might be the production of nontrivial product compounds and the stimulation of chemical reactions by electronically excited molecules. The latter might be important for large molecules and biological materials.

The possibility of using an inverted population obtained in this way requires further study in view of its oscillatory character. The problem of building a coherent frequency converter (see Fig. 6) is also nontrivial. If the transition is highly probable then under certain conditions a multiple of the laser frequency, $\hbar\omega' = m\hbar\omega$, may be obtained.

These applications are a long way from exhausting all the applied aspects of this radiative mechanism for exciting multiatomic molecules.

6. A Comparison of the Basic Interactions of Laser Radiation with Atoms and Molecules

The discussion of the previous sections shows that under certain conditions the radiative coherent mechanism is dominant in the case of multiatomic molecules. These are not isolated results. The coherent mechanism is very general and may make an important contribution to the ionization of atoms (and diatomic molecules) at high intensities. Here we shall discuss these questions purely qualitatively. Usually [11] atomic ionization in a laser field is regarded as the result of two ionization mechanisms. The initial electrons are removed from the atoms due to multiphoton photoelectric effect. Then the electrons gain energy in the laser field during collisions due to inverse bremsstrahlung and an electron avalanche develops.

In addition, it should be noted that multiphoton processes have thresholds, and at fields corresponding to the threshold (10^5 V/cm) atomic levels are strongly "broadened." Thus, for the upper levels where the distance between the levels is relatively small, "broadening" may be sufficient for the coherent mechanism to be realized. The required threshold intensities may differ greatly depending on the structure of the terms. With fields of 10^7 V/cm the "broadening" is insufficient for operation of the coherent mechanism in the case of a single electron in an outer shell of an atom. The situation changes with two or more electrons. The estimates that have been made (taking into account the change in the transition probability in the laser field) do not contradict the experimental data given in [11].

Below we list the number of photons up to the ionization limit n for a number of elements along with the experimentally obtained number n_0 (the energy of a single laser photon is 1.18 eV).

Element	Xe	Kr	Ar	Ne	Na	K	H_2	He II
n	11	12	14	19	5	4	14	3
n_0	9	9	10	13	5	4	8	3

The experimental values of n and n_0 given here have been rounded off since they were obtained for different fields ($\sim 10^7$ V/cm), which has an additional effect on their magnitude.

The difference $n - n_0$ corresponds to the limits of operation of the radiative coherent mechanism since estimates of its rate yield values much greater than that for the multiphoton process. As soon as the electrons reach the upper levels corresponding to n_0 due to multiphoton excitation, the ionization due to the coherent mechanism may be regarded as instantaneous. We note that the problems associated with rapid ionization from upper excited levels have been discussed previously (see [11, 12, 26, 27]).

Therefore, in general the analysis of atomic ionization in a powerful laser field requires joint consideration and calculation of three basic mechanisms: (a) multiphoton excitation of relatively high-lying levels; (b) radiative coherent ionization from these levels; and (c) collisional energy gain by the electrons and the production of electron avalanches.

A similar situation occurs for ionization of diatomic molecules. Here the coherent mechanism operates between the dissociation and ionization limits with transitions occurring between different electronic states of the molecules.

In the case of multiatomic molecules at moderate densities the radiative and collisional excitation, dissociation, and ionization mechanisms must be considered together. An experimental study of molecular excitation under these conditions was made in [13]. It was found that there is a threshold for the luminescence intensity I_{lum} of the fragments formed during dissociation. There

$$I_{\text{lum}} \sim I^{m}_{\text{pump}},$$

and the exponent on the pumping intensity, m, depended on the pressure ($m = 10^5 p^{-1.7}$, with p in atmospheres). A delay in the onset of dissociation was also observed.

An analysis of radiative–collisional processes, as in the case of purely collisional processes [9, 10], must be based on the relationships among the characteristic times. The following characteristic times may be pointed out: $\tau' = 1/\mathcal{E}d_3$, the radiative kinetic time; τ_{VV}, the vibrational exchange time; τ_{pulse}, the pulse duration; and τ_{D}, the time for dissociation (or ionization) to occur. For $\tau_{\text{pulse}} \ll \tau_{\text{VV}}$ and $\tau' \ll \tau_{\text{D}}$ the above coherent mechanism is valid. For $\tau' \ll \tau_{\text{D}}$ and $\tau' \ll \tau_{\text{VV}} \ll \tau_{\text{pulse}}$ we might expect the "fluctuating" populations picture but without repetition over a long period (i.e., a weakly stochastic coherent mechanism). For $\tau_{\text{VV}} \lessgtr \tau'$ the model of an N-level (with N ~ 30–40) oscillator ceases to be valid and effective resorption (dissipation) of the populations takes place over all degrees of freedom of a multiatomic molecule.

The case $\tau_{\text{VV}} \gg \tau'$ and $\tau_{\text{pulse}} \leq \tau_{\text{VV}} \ll \tau_{\text{D}}$ is of special interest. Then the populations produced as a result of radiative excitation are the initial populations for formation of the distributions due to molecular collisions. Thus, a Treanor distribution [10, 28] is formed over the ν_3 vibration in the case of excitation of the BCl_3 molecule. On the basis of this [9, 29] it is possible to compute the dissociation rate and its time variation. In general the role of quasiequilibrium distributions (see [30]) can be very important in the analysis of the radiative–collisional mechanism.

In conclusion, we emphasize that the resonant coherent mechanism is very general and is involved in a wide range of phenomena. Our discussion has demonstrated the fundamental role of the spectrum of the transition dipole moments in this mechanism. The set of dipole moments (along with the power and the pulse length) determines the entire population pattern in every concrete case as well as the rate of radiative dissociation (or ionization) of the molecules. Although this set is still little known, one may expect with high probability that the process of populating the higher levels will be very effective both in the excitation of degenerate vibrations (due to nonrandom Fermi resonance) and with other modes when the random Fermi resonance occurs. However, for a quantitative solution (and in a number of cases, even a qualitative understanding) of specific problems in the area, the first task is to study the dipole moment spectrum.

LITERATURE CITED

1. F. V. Bunkin and A. M. Prokhorov, Zh. Éksp. Teor. Fiz., 46:1090 (1964).
2. L. V. Keldysh, Zh. Éksp. Teor. Fiz., 47:1945 (1964).
3. G. S. Voronov and N. B. Delone, Zh. Éksp. Teor. Fiz., 50:78 (1966).
4. H. B. Bebb and A. Gold, Phys. Rev., 143:1 (1966).
5. G. A. Askar'yan, Zh. Éksp. Teor. Fiz., 46:403 (1964); 48:666 (1965).
6. N. D. Artamonova, V. T. Platonenko, and R. V. Khokhlov, Zh. Éksp. Teor. Fiz., 58:2195 (1970).
7. Yu. V. Afanas'ev, E. M. Belenov, and O. N. Krokhin, Zh. Éksp. Teor. Fiz., 56:256 (1969).
8. N. V. Karlov, Yu. V. Konev, and A. M. Prokhorov, Pis'ma Zh. Éksp. Teor. Fiz., 11:20 (1970).
9. B. F. Gordiets, Sh. S. Mamedov, and L. A. Shelepin, Zh. Éksp. Teor. Fiz., 67:1287 (1974).
10. B. F. Gordiets, A. I. Osipov, E. V. Stupochenko, and L. A. Shelepin, Usp. Fiz. Nauk, 108:655 (1972).
11. Yu. P. Raizer, Laser Sparks and the Propagation of Discharges [in Russian], Nauka, Moscow (1974).
12. N. B. Delone, Usp. Fiz. Nauk, 115:361 (1975).
13. N. V. Karlov, Yu. N. Petrov, A. M. Prokhorov, and O. M. Stel'makh, Pis'ma Zh. Éksp. Teor. Fiz., 11:220 (1970).
14. N. R. Isenor, V. Merchant, R. S. Hallsworth, and M. C. Richardson, Can. J. Phys., 51:1281 (1973).
15. R. R. Ambartsumyan, V. S. Dolzhikov, V. S. Letokhov, E. A. Ryakov, and N. V. Chekalin, Abstracts of talks at the VIIth All-Union Conference on Coherent and Nonlinear Optics, Izd. MGU (1974), p. 486.
16. R. R. Ambartsumyan, V. S. Letokhov, E. A. Ryabov, and N. V. Chekalin, Pis'ma Zh. Éksp. Teor. Fiz., 20:597 (1974).
17. A. N. Oraevskii and V. A. Savva, Kratk. Soobshch. Fiz., No. 7, p. 50 (1970).
18. V. A. Kravchenko and A. S. Prostnev, Dokl. Akad. Nauk SSSR, 211:73 (1973).
19. S. S. Alimpiev and N. V. Karlov, Zh. Éksp. Teor. Fiz., 63:482 (1972); 66:542 (1974).
20. L. Wolniewicz and T. Kowalski, Chem. Phys. Lett., 18:55 (1973).
21. B. V. Gnedenko, A Course in Probability Theory [in Russian], GITTL, Moscow (1954).
22. V. Ya. Yakubovich and V. M. Strazhinskii, Linear Differential Equations with Periodic Coefficients [in Russian], Nauka, Moscow (1972).
23. V. M. Akulin, S. S. Alimpiev, N. V. Karlov, and L. A. Shelepin, Zh. Éksp. Teor. Fiz., 69:836 (1975).
24. R. R. Ambartsumyan, V. S. Dolzhikov, V. A. Letokhov, E. A. Ryabov, and N. V. Chekalin, Zh. Éksp. Teor. Fiz., 69:72 (1975).

25. V. M. Akulin, S. S. Alimpiev, N. V. Karlov, N. A. Karpov, Yu. N. Petrov, A. M. Prokhorov, and L. A. Shelepin, Pis'ma Zh. Éksp. Teor. Fiz., 22:100 (1975).
26. G. S. Voronov, Zh. Éksp. Teor. Fiz., 51:1496 (1966).
27. A. P. Kotova and M. V. Terent'ev, Zh. Éksp. Teor. Fiz., 52:732 (1967).
28. C. E. Treanor, J. W. Rich, and R. G. Rehm, J. Chem. Phys., 48:1798 (1968).
29. B. F. Gordiets, A. I. Osipov, and L. A. Shelepin, Zh. Éksp. Teor. Fiz., 61:562 (1971).
30. L. A. Shelepin, Trudy FIAN, 83:3 (1975); S. A. Reshetnyak, Trudy FIAN, 83:193 (1975).

GPSR Compliance

The European Union's (EU) General Product Safety Regulation (GPSR) is a set of rules that requires consumer products to be safe and our obligations to ensure this.

If you have any concerns about our products, you can contact us on ProductSafety@springernature.com

In case Publisher is established outside the EU, the EU authorized representative is:

Springer Nature Customer Service Center GmbH
Europaplatz 3
69115 Heidelberg, Germany

Batch number: 10845493

Printed by Printforce, the Netherlands